THE
GREAT WHITE
SHARK
HANDBOOK

ISBN-13: 978-1-60433-771-6
ISBN-10: 1-60433-771-0

This book may be ordered by mail from the publisher. Please include $5.99 for postage and handling. Please support your local bookseller first!

Books published by Cider Mill Press Book Publishers are available at special discounts for bulk purchases in the United States by corporations, institutions, and other organizations. For more information, please contact the publisher.

Cider Mill Press Book Publishers
"Where good books are ready for press"
501 Nelson Place
Nashville, Tennessee 37214

cidermillpress.com

Typography: Adobe Text Pro, PF Venue

Photography Credits: Page 7 courtesy of Wilson Skomal; Page 17 courtesy of Greg Amptman and undeseadiscoveries.com. Pages 21, 22, 25, 36, 39, 45, 49, 54, and 59 courtesy of Victoria Wilchinsky; Page 74 courtesy of Tobey Curtis; Pages 77 and 131 courtesy of NOAA/NMFS; Page 79 courtesy of Carey et al., 1982; Pages 80, 152, and 194 courtesy of Brian Hanson; Pages 88, 106 (left), 107 (bottom), 108 (left), 114 (left), 149, 173, 187, and 191 (right) courtesy of Wayne Davis; Pages 100, 102 (top), and 116 courtesy of John Chisholm; Page 102 (bottom) courtesy of Pam King; Pages 108 (right) and 111 courtesy of Chris Fallows; Pages 112 and 113 courtesy of Woods Hole Oceanographic Institution; Page 140 courtesy of Wes Pratt; Pages 143 and 144 courtesy of Jon Dodd; Page 209 courtesy of the Cuttyhunk Historical Society. Pages 8, 12, 30, 57, 70, 99, 142, 162, 174, 178, 181, 195, 202, 211, and 212 used under official license from Shutterstock. All other photos courtesy of the author.

Printed in Malaysia

24 25 26 27 28 OFF 5 4 3 2 1
First Edition

THE
GREAT WHITE
SHARK
HANDBOOK

THE DEFINITIVE GUIDE
TO THE MOST EXTRAORDINARY
SHARK IN THE SEA

GREG SKOMAL

CIDER MILL
PRESS

BOOK
PUBLISHERS

CONTENTS

INTRODUCTION

I grew up in southwestern Connecticut, not too far from New York City. Like most kids, I was infatuated with sharks. I thought they were really cool looking with their sleek bodies, sharp fins, and big teeth. The fact that they were considered dangerous also fascinated me. Where I lived, the closest ocean was the Atlantic via Long Island Sound, which was not the most pristine body of water back in the 1970s. We went to the beach a bit, but the water was dark and frightening to me back then, so I didn't do much exploring. Instead, I immersed myself in any television program that showed the ocean, which was mostly crystal-clear tropical waters teeming with life, including sharks. My favorite was the *Undersea World of Jacques Cousteau*. To me and many others back then, Jacques was a pioneer in underwater exploration. I couldn't get enough, especially when he showed sharks, which was almost every episode. That seed would ultimately grow into my love for the ocean and its inhabitants.

In the early 1970s, my parents took our family to the Caribbean, where I immersed myself in warm tropical waters and experienced the diversity of fishes in real life. I snorkeled for hours on end pretending I was an explorer, always looking for sharks,

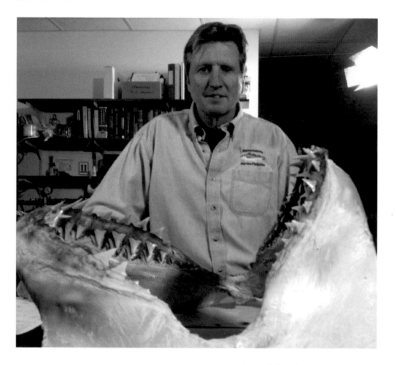

Author and shark biologist, Greg Skomal

but never seeing one. Back in Connecticut, I set up an aquarium so I could keep my fascination with fishes alive. I was obsessed. In 1975, my preoccupation with sharks exploded when I went to the local movie theater and saw *Jaws*. Sure, I was horrified by the man-eating behavior of the shark but, more importantly, I was drawn to the shark biologist in the movie, Matt Hooper—I wanted his job. For many people, the movie's biggest impact was scaring people out of the water, but it was the opposite for me.

Over the course of the next twenty years, I went to the University of Rhode Island to study marine biology and obtained Bachelor of Science and Master of Science degrees; I volunteered, then worked with top shark researchers at the world-renowned Cooperative Shark Tagging Program, which was followed by a job as the shark biologist for the Massachusetts Division of Marine Fisheries. Get this, I was stationed on Martha's Vineyard where *Jaws* was filmed. Over the last thirty-six years, I've had the opportunity to study the ecology, life history, and physiology of sharks, not only off the coast of New England but in many locations throughout the world. But like many shark biologists, I've always wanted to study the great white shark.

In the western North Atlantic, the great white shark, which is what scientists generally call the white shark, was impacted by overfishing in the 1970s and 1980s. In addition, seal populations were driven to the brink of extinction during the nineteenth and twentieth centuries. White sharks like to feed on seals close to shore, so when the seals were extirpated, the sharks spread out over the broader oceanic waters of the Atlantic. With a low population, white sharks became very difficult to predictably find, so scientists relied on the incidental capture of specimens by fishermen to learn about the species. As a young shark biologist in the 1980s, my only experience with white sharks was when I dissected some of these dead specimens to examine various aspects of their life history. As you will see in this book, you can learn a lot from a dead shark, but not much about its behavior, movements, and ecology. The dead sharks only brought me so far in my quest for understanding—I wanted to study live white sharks.

With the passage of the Marine Mammal Protection Act in 1972, and the prohibition on landing white sharks in 1997, both populations began to rebound. In the early 2000s, this conservation success story started to manifest itself off the coast of Cape Cod, Massachusetts. It started with the frequent discovery of injured seals and carcasses on the beaches, but by 2009 we began to observe white sharks hunting gray seals very close to shore along the eastward facing beaches of the Cape. And so began my research on live white sharks in the Atlantic. A dozen years later, the partnership of my agency, the Division of Marine Fisheries, with the Atlantic White Shark Conservancy has resulted in the most comprehensive and successful white shark research and education program in the Atlantic Ocean.

Although I have seen more than a thousand white sharks over the course of my career, I never get tired of searching for and finding the next one. I am indeed living a dream studying these animals. The more I learn, the more I realize how much we don't know. My guess is that you share my fascination with this species. Arguably, the white shark is the most charismatic and, at the same time, the most vilified species on the planet. It is more frequently portrayed as a killer than any other animal, other than humans, yet people are still drawn to it. All forms of media—newspapers, websites, social media, films—know that a man-eating white shark sells, and sells big. So, what you typically see of this species is more likely to scare you than cultivate your fascination. I want to set the record straight.

In this book, you will learn about the real white shark, not the one you typically see in the media. I'm going to tell you everything I know about this species, not only what I have personally experienced, but what I have learned from the work of other researchers around the world. You will read about its evolution, anatomy, distribution, movements and migrations, reproduction, feeding ecology, growth rates, and longevity. I will cover where you can see white sharks in their natural environment, the status of their populations, and white shark conservation. And I'm not going to shy away from shark attacks—I will tell you what we know, what and where the risks are, and potential solutions. If you want to be a white shark biologist or are simply just fascinated by the species, then this is the place to start. Jump in the water with me!

1
EVOLUTION

The great white shark is one of the most advanced, or highly evolved, species of sharks. As you will see in the next few chapters, this species has been finely honed through evolution to be a top predator in all the world's oceans. But none of this happened overnight. The evolution of this and all shark species started more than 450 million years ago.

CLASSIFICATION

Before we focus on the origins of the white shark, it is important that you get a good understanding where this species fits into the grand scheme of life on earth. Scientists classify all living things, and sharks are no exception. To fully appreciate the white shark, let's first review the taxonomy, or classification, of all sharks. Like all fishes, sharks are classified using similar attributes, including a variety of physical characteristics like skeletal shape and number of vertebrae. In recent times, genetic analyses have allowed researchers to examine the evolution and relationships among living as well as fossilized sharks. While there is always ongoing debate and discussion about how sharks are related and named, I'm going to present the most widely accepted classification.

TAXONOMY

Humans speak many different languages, so the naming of living creatures can be a problem. For example, "white shark" in English is "tiburon blanco" in Spanish. To get around the confusion associated with language differences, a Latin-based scientific naming scheme, now called *binomial nomenclature* ("two names"), was established by Carl Linnaeus in 1735. The Linnaean system classifies all living things in a hierarchy. At its lowest level, an individual type of creature is given a scientific name composed of two parts: the genus and species. Very closely related species belong to the same genus. For example, the genus *Carcharhinus* contains at least thirty-four closely related species, including the bull shark, *Carcharhinus leucas*, and the dusky shark, *Carcharhinus obscurus*. If you look at these two fish, you can see that they are extremely similar. This classification scheme continues as you group similar genera (plural for genus) into families. Then you group similar families into orders, similar orders into classes, and similar classes into phyla (the singular is phylum). All the different phyla make up the kingdom that we call Animalia for the animals. There are also subdivisions of these categories, like suborder and infraclass. The basis of establishing these groups rests on the *phylogeny*, or ancestral lineage, of these animals. Therefore, taking this into consideration, families and orders of sharks are grouped in such a way as to imply that they evolved from similar common ancestors.

WHAT IS A SHARK?

A common misconception is that sharks are not fish. Sharks belong to a class of fishes known as Chondrichthyes, which is more commonly referred to as the cartilaginous fishes. The chondrichthyan fishes are vertebrates, like mammals, reptiles, birds, and amphibians, belonging to the subphylum Vertebrata, which belongs in the phylum Chordata. All the chordates are animals, so, as opposed to plants, these critters belong to the Kingdom Animalia. The class Chondrichthyes also includes skates, rays, chimaeras, and a huge number of extinct species, extending back over 450 million years. Although there are a whole bunch of technical ways to distinguish chondrichthyan fishes from all others, the most conspicuous characteristics include an internal skeleton composed of cartilage, toothlike scales called *denticles*, modified pelvic fins called *claspers* (males only) used for internal fertilization, and teeth in replicating rows. There are more than 1,200 living species of chondrichthyan fishes, but this number seems to keep growing each year, as more and more species are described.

Since sharks, rays, and chimaeras look very different, the class Chondrichthyes is subdivided into the subclasses Elasmobranchii, which includes the sharks and rays, and Holocephali, the chimaeras. The Holocephali comprise fifty-six living species of chimaeras, or ratfishes. These deepwater fish reach up to 6 feet in length and

Chimaera or ratfish

can best be distinguished from sharks and rays by their single gill covers, smooth skin, enormous spines, fixed upper jaws, and large tooth plates instead of teeth. The balance of living chondrichthyans, the Elasmobranchii, includes the sharks and their closest relatives, the skates and rays. To differentiate between the sharks and rays, the elasmobranchs are broken into the Superorder Batoidea, comprising the skates and rays, and the Superorder Selachimorpha, which are the sharks. The more than 530 species of living sharks are easily distinguished

from their raylike relatives by their general torpedo shape. In contrast, most of the 689 species of skates, rays, guitarfishes, and sawfishes look very much like flattened sharks, with broad, winglike fins attached to their heads.

But wait, there is still more subdivision. Not all sharks look, live, or act alike, so science has grouped them based on common features into nine orders:

Carcharhiniformes	**Pristiophoriformes**	**Heterodontiformes**
Lamniformes	**Orectolobiformes**	**Echinorhiniformes**
Squaliformes	**Hexanchiformes**	**Squatiniformes**

This blue shark belongs to the order Carcharhiniformes.

Now let's look specifically at the great white shark. First, most people call this species the "great white shark," which is fine, but the scientific community decided many years ago to drop the "great" and simply call it the "white shark." So, for most of this book, I will use the common name "white shark." Of course, you are welcome to call it a great white shark—I think they are pretty great as well.

The white shark has the genus and species name *Carcharodon carcharias*. There is no other living shark species in the genus *Carcharodon*, so that tells you this shark is unique. But it does have cousins in the same family, which is called Lamnidae. Based on genetic and other analyses, the lamnid sharks (in the family Lamnidae) are similar to sharks in seven other living shark families and five extinct shark families. They are all grouped together in the order Lamniformes. So, here is a summary of the white shark's classification.

The mako shark is a close relative of the white shark.

KINGDOM **Animalia**

PHYLUM **Chordata**

SUBPHYLUM **Vertebrata**

SUPERCLASS **Gnathostomata**

CLASS **Chondrichthyes**

SUBCLASS **Elasmobranchii**

SUPERORDER **Selachimorpha**

ORDER **Lamniformes**

FAMILY **Lamnidae**

GENUS *Carcharodon*

SPECIES *carcharias*

By tracing this hierarchy, we can follow the evolution of the white shark and its closest relatives.

The white shark belongs to the family Lamnidae.

ORIGINS AND ANCESTORS

Sharks have prowled the world's waters for the last 450 million years. Compare that to modern humans who have only been around for about 200,000 years! It is remarkable to think that some of the shark species, but not the white shark, that swim in our oceans today were around when dinosaurs roamed the earth. What is even more amazing is that the dinosaurs became extinct about 65 million years ago, but the sharks persisted and still dominate the watery world. Ironically, after all this time, the only threat to sharks is humans.

The anatomical feature that binds all chondrichthyan fishes, namely cartilage, has also rendered their origins and ancestry very much a mystery. Unlike the bones of other fish species, cartilage does not preserve well in the fossil record. Therefore, with few exceptions, the trail of shark evolution is marked only by the fossilized remains of dermal denticles, teeth, and vertebrae. To really get a sense of where sharks fit into the fossil record, we must first put life on Earth in geologic perspective—this is done with a geologic time scale. Given our propensity to classify, organize, and count, it's no surprise that we have divided the 4.5 billion-year history of our planet into eons, eras, periods, and epochs.

The Precambrian eons span about 88 percent of Earth's history, up to 539 million years ago. Life on Earth originated during this time and has become increasingly complex since. The ancient relatives of sharks first appeared in the fossil record about 450 million years ago during the Silurian period, but the fossilized teeth of what may be considered true sharks didn't appear until the Late Devonian, about 380 million years ago. Based on the size of these teeth, the early sharks were small, only about a foot long. It was during the Devonian period (about 419 to 359 million years ago), often referred to as the "Age of Fishes," that the number of fish species exploded, and sharks were no exception.

Cladoselache

Among these earliest sharks was *Cladoselache*, which was described from teeth, vertebrae, and rare fossilized impressions of skin, kidneys, and muscle. With its sharklike body, sharklike fins, and sharklike large eyes, *Cladoselache* was likely about 6 feet long and a fast-swimming, predatory shark. The Carboniferous period, about 359 to 299 million years ago, became what we now call the "Golden Age of Sharks." Although *Cladoselache* had come and gone by then, other species of sharks radiated from common ancestors. Among them were the earliest hybodonts, a group of sharks with fins, claspers, spines, and conical teeth that more closely resembled those of the sharks that roam our oceans today. The rapid proliferation of sharks during this period also resulted in the development of adaptations and specializations that look very bizarre, including the odd looking *Stethacanthus* with a crown of tubercle-type teeth on its head. Shark diversity during the Carboniferous period was staggering, and the number of shark families at that time actually outnumbered those alive today.

Stethacanthus

However, by the close of the next period, the Permian, about 251 million years ago, most of these sharks were eradicated by the largest extinction event of all geological time. A domino-type series of climatic and geochemical changes caused by extensive volcanic activity spanning thousands of years resulted in the loss of not only sharks, but as many as 95 percent of all marine species. Nonetheless, the adaptability and success of the shark design would persevere through this global catastrophe as a couple of shark groups emerged, quite literally, from the ashes to seed another burst in shark evolution. These groups included the hybodonts, which survived and roamed the seas for another 180 million years. The other, the xenacanth sharks, were small- to medium-sized sharks (3 to 6 feet long) that lived in freshwater and looked more like eels, with a single dorsal fin that wrapped around the tail and a conspicuous spine on the top of the head.

Although the Jurassic period (201 to 145 million years ago) is defined as the "Age of Dinosaurs," it also marked the appearance of the first modern sharks. At one time, it was thought that the hybodonts gave rise to the lineage of sharks that we see today, but this does not seem to be the case. Instead, the hybodonts were being replaced by species with more flexible protrusible jaws, a characteristic that would chill the bones of many a human millions of years later. The species that emerged during the Jurassic and the following period, the Cretaceous (145 to 66 million years ago), are now typically referred to as the *neoselachians* or the "new sharks." Although not as prolific as the Golden Age of Sharks 200 million years earlier, these periods mark the emergence of the major lineages of almost all of the sharks that we see today with incredibly diverse forms and lifestyles. To compete with the marine dinosaurs, the new sharks had to be streamlined, fast, large (some longer than 20 feet), and loaded with a complement of senses to assist with prey detection and capture. The oldest of the neoselachians are the lineages that gave rise to our modern-day cow (Hexanchiformes) and bullhead (Heterodontiformes) sharks about 95 to 150 million years ago. It was also during this time that the skates and rays appeared in the fossil record; some of these ultimately invaded and still reside in freshwater. The appearance of teeth that looked remarkably like those of our modern lamniform sharks, the white shark and its cousins, marks the emergence of prototypes for these modern-day species as well.

ENTER THE WHITE SHARK

It is no secret to anyone who looks out their window that the Age of Dinosaurs did not persist to modern times. Indeed, another catastrophic extinction occurred at the end of the Cretaceous period about 65 million years ago. Thought to be caused by an asteroid impact, the Cretaceous-Tertiary extinction event resulted in the loss of all the dinosaurs, with the exception of those giving rise to our modern birds. Although the declining hybodonts met their demise along with the dinosaurs, remarkably about 80 percent of the Chondrichthyan families survived the event. With the loss of the dinosaurs both above and below the surface, mammals flourished in both places, as access to vacated habitat was now possible, and the ancestors of marine mammals, like whales and dolphins, returned to the sea. These creatures provided ample fodder to fuel the development of larger predatory sharks like the white shark.

First and foremost, the white shark did not evolve from the megalodon shark, *Otodus megalodon*. Considered the largest predatory fish to have ever lived, the megalodon shark lived from about 23 to 3.5 million years ago. It was the last species in the lineage of megatoothed sharks of the family Otodontidae, although there is still some residual debate over whether this shark should be placed in the genus *Carcharocles*. At an estimated length of more than 50 feet and weight of 60 tons, this shark easily could have consumed the largest white shark, which stands at 20 feet. The megalodon has the largest shark teeth (up to 6 inches) ever examined, and they are remarkably similar in form to the triangular, serrated teeth of the

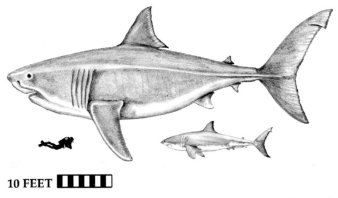

10 FEET

The megalodon shark was 20 times heavier than the white shark.

Megalodon tooth

white shark. For this reason, many researchers hypothesized that the megalodon shares a common lineage with the white shark. However, recent studies suggest that megalodon evolution paralleled that of the modern white shark in a separate lineage. Megalodon's fossils are found worldwide and predominantly in coastal regions side by side with those of large sea mammals, their likely prey. At 6 feet across, the mouth of this shark was immense and more than capable of dispatching large creatures, even whales. Exactly what precipitated the extinction of megalodon remains a mystery, but it may have been associated with changes in prey distribution, climate change, the collapse of prey populations, and even resource competition with the white shark. Although there are some who feel that a relict population of megalodon sharks exists deep in the world's oceans, this unfortunately can only be characterized as wishful thinking.

White sharks are actually distant relatives of the mako shark. The best available evidence indicates that the white shark lineage shared a common ancestor with the mako lineage. This means that the broad, serrated teeth of the white shark have evolved from an ancestral mako-like species with pointed, non-serrated teeth, which still persist in living makos. Although the evolutionary scenario appears to change with every new fossil discovery, this transition in tooth structure can be seen in a series of extinct species. The common mako-like species might have been *Isurolamna inflata*, which

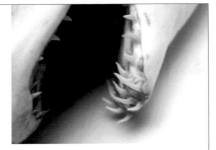

Ancestors of the white shark had teeth similar to those of this mako.

had smooth, pointed teeth and lived between 65 to 55 million years ago. Next came *Macrorhizodus praecursor*, which also had smooth teeth, but they were a bit broader. This species gave rise to the broad-toothed mako, *Carcharodon hastalis* or *Cosmopolitodus hastalis* (depending on who you ask), which lived about 35 to 1 million years ago and had even broader teeth. It is believed that *Carcharodon hubbelli*, which emerged about 8 to 6 million years ago, is an intermediate species between *hastalis* and today's great white shark. It has very similar teeth that are serrated, yet not as broad as those in the white shark. Based on the fossil record, it is generally thought that the modern white shark, *Carcharodon carcharias*, emerged in the Pacific Ocean during the Late Miocene and Early Pliocene about 5 million

White shark tooth

years ago. Like its ancestors, the species roaming the ocean today is a top predator in the marine ecosystem.

Although the white shark is categorized as a single species that is now distributed throughout the world, there are very fine genetic differences between the existing populations. The explosion in molecular research on white shark DNA in recent years, including the sequencing of its complete genetic material (*genome*), has allowed researchers to also speculate on its origins. Based on a number of these analyses conducted by various researchers, distinct genetic differences have been found between all the global

Today's white shark has been around for about 5 million years.

white shark populations. Genetic material from females indicates that the species might be *philopatric*, which means that they return to where they were born to give birth to their young. This has resulted in reproductively isolated populations around the world. However, white shark populations can still be clustered into two main lineages based on genetic differences. Those sampled in the Northwest Atlantic cluster with white sharks sampled off South Africa, and white sharks sampled in the Southwest Pacific (Australia and New Zealand) cluster with those sampled in the Northeast Pacific (California) and the Mediterranean Sea.

These relationships have allowed researchers to speculate on how white sharks dispersed throughout the world. Remember, all available evidence indicates that the species originated in the Pacific Ocean in the region of Australia about 5 million years ago. The most recent hypothesis indicates that these ancient white sharks dispersed by moving westward to South Africa as well as eastward to the Northeast Pacific. This latter group of sharks continued to move east between North and South America through the Central American Seaway because the isthmus of Panama did not exist. By about 3 million years ago, the species spread across the North Atlantic and populated the Mediterranean Sea. Under this scenario, the Northwest Atlantic population is thought to have originated more recently from individuals that moved north and west from South Africa. However, as is the case with most evolutionary stories, the white shark's is likely to change with the discovery of new fossils. Stay tuned.

2
ANATOMY

Although white sharks are at the very top of the food web, survival in the ocean is very different from living on land. Water is a thick medium, about 800 times denser than air, so just moving requires a lot of work. Life in the ocean is not uniformly distributed, so if you are trying to make a living, you need to be at the right place at the right time. This oceanic patchiness forces many marine animals to move seasonally to and from productive areas, while of course keeping within their physiological limits. These movements must also coincide with reproductive timing. As a result, white sharks move great distances to feed and to reproduce, while other species, like tropical reef fishes, stay put to do both. Survival in the ocean, therefore, hinges on energetic tradeoffs between growth, foraging, predator avoidance, and reproduction in a dense medium and an environment with scattered resources. Moreover, that same medium, namely seawater, is loaded with salt, so the white shark and all other fishes must deal with the physiological fact that their bodies are constantly losing freshwater to the environment. Add to this the odd properties of light and sound in water, and the need to capture both in order to see and to hear, and you have a world where making a living is not easy, even for the white shark.

There are several major anatomical features of sharks, in general, that make them unique among the fishes and you can think of the white shark as being on the high end of all shark species, as a Tesla or a Porsche stands in regard to other automobiles. Having a solid understanding of these features not only aids in classification and identification, but it also helps researchers investigate how white sharks respond to and live in their environment. Exactly how the white shark survives in the marine world is dictated by its morphology, physiology, and sensory systems. In the previous chapter, we learned that the white shark has changed little over millions of years. Why? Because those attributes that define this species have allowed it to survive and thrive in almost all parts of the world from the shallowest seas to the deepest oceans and from the tropics to cool temperate zones.

As we learned in the last chapter, there are nine orders of sharks, each containing families and species that are thought to be related phylogenetically, that is, each has the same evolutionary lineage. As explained in the previous chapter, the white shark belongs to the order Lamniformes with its closest relatives, which have very similar characteristics. On the outside, body shape, tooth structure, fin size and placement, and even scale configuration are the most obvious features. If you could look on the inside, you would see variations in muscle and skeletal structure, reproductive and digestive systems, and nervous and sensory system arrangement, just to name a few. These features are different in the Lamniform species, but consistent in every white shark, although there are subtle differences between individuals. Like all living species, the white shark has adapted to meet the challenges of living in the ocean. Most of these adaptations are associated with balancing energetic losses and gains. As you will see in this chapter, the white shark's anatomy is characteristic of a highly specialized, fast-swimming apex predator at the top of the food web.

MORPHOLOGY

The general form and structure, called morphology, of a white shark includes several major features that differentiate them and other shark species from all other fishes. This unique morphology includes an internal skeleton composed of cartilage, tooth-like scales called denticles, the lack of any kind of swim bladder, modified pelvic fins called claspers used for internal fertilization, and teeth in replicating rows.

BODY SHAPE AND FINS

The first thing that anyone notices about the white shark is its streamlined torpedo-shaped (referred to as *fusiform*) body, subterminal mouth (under its head), fins, and large size. The white shark, like all fishes, has adapted to living in a very dense medium. Every sweep of its tail is work, which requires energy, and the white shark must continuously swim to breathe. So, the white shark's body shape, fin and tail shape, and even scale alignment and shape, are designed to allow water to flow over it, thereby reducing drag and increasing energy efficiency. For comparison, look at the efficient hydrodynamic shapes of a torpedo and a submarine. They are not box-shaped for a reason!

The white shark's dorsal fin helps to stabilize the shark during fast swimming.

Other than their impressive jaws, the most famous part of a white shark is its dorsal fin. All of us have seen those movies when somebody screams and points to a fin neatly slicing the surface of the water—the telltale sign of a shark. Of course, Hollywood doesn't particularly care to know that white sharks spend a lot of time in the deep and don't come to the surface as frequently as portrayed in movies. Aside from scaring people, the dorsal fin of the white shark is unique in its design. White sharks have densely concentrated bundles of collagenous dermal fibers that extend from the body well into the dorsal fin just under the skin. These fibers act to stiffen the fin during fast swimming, thereby stabilizing the shark and allowing it to swim straight when the potential to spin in the water is greatest.

In addition to the dorsal fin, white sharks have a much smaller second dorsal, a caudal or tail fin, a small anal fin, and

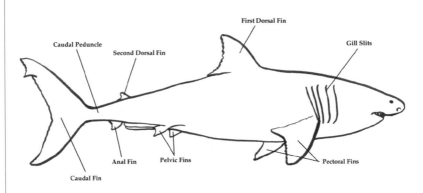

The fins of the white shark.

two sets of paired fins, the pectoral and pelvic fins. The pectoral, pelvic, dorsal, and anal fins help the shark maneuver, rise and fall in the water column, turn, and swim straight. Unlike the fins in most bony fish species, which are quite maneuverable, white shark fins are fixed in place and remain erect, but muscles in the fins do allow for subtle movements. These fins are supported by cartilaginous radials and keratin-based supportive elements called *ceratotrichia*, which look a bit like dry

spaghetti running up through the fin. The ceratotrichia are also the main ingredient of shark fin soup, a delicacy that has driven some shark populations to the brink of extinction.

The caudal fin is the white shark's propeller, providing the thrust and power needed to drive the shark through the water. Shark tails, in general, are much different from those of other fishes in that the vertebral column passes up into the upper lobe of the tail, allowing for the power to be generated along the entire axis of the body. This results in a heterocercal tail, in which the upper lobe is generally larger and longer than the lower lobe. In the great white and other lamnid sharks, the lower lobe is almost equal in length to the upper lobe, resulting in a lunate tail, which is characteristic of fast-moving fishes like tunas. The narrowing of the body where it meets the tail is referred to as the caudal peduncle. White sharks and their cousins have a very strong keel on either side of the caudal peduncle. This allows all the force generated by the shark's muscles to be translated efficiently to the tail through the caudal peduncle.

Sharks, in general, can be grouped generally by their basic body and fin designs, depending on their lifestyle. These groups range from the sedentary, bottom-dwelling species like angelsharks, to mid-sized, moderate swimmers like tiger sharks, to highly specialized, fast-swimming open-ocean species like the shortfin mako. The white shark belongs to this more specialized group of

White sharks and their relatives have very strong caudal keels.

fast-swimming species, which include other lamnid sharks, tunas, dolphins, and the extinct ichthyosaurs. This lift-based mode of swimming is referred to as *thunniform* because it was first described in tunas of the genus *Thunnus*. Species that use this mode of locomotion have a conical head, a large deep body, large pectoral fins, a narrow caudal peduncle with keels, and a symmetrical lunate caudal fin. This is the most efficient body design for moving quickly through the water and their small second dorsal, anal, and pelvic fins help to reduce drag.

Bluefin tuna

SKIN

Sharks have tiny tooth-like scales called placoid scales or dermal denticles covering their bodies, all pointing toward the tail. Denticles are composed of enameloid (enamel-like tissue), dentine, and a central pulp cavity. When viewed from the side, you can see a base that inserts into the skin, and a narrow stalk connected to a broad crown with ridges and cusps pointing backward. This orientation results in the smooth feel of a shark's skin when stroked toward the tail but the classic sandpaper texture when stroked toward the head—something you should know when petting a shark. In fact, the skin is so rough that it has been used as sandpaper. On more than one occasion, I have been abraded by the skin of a shark while handling it and it's no different than swiping sandpaper over your skin, which can be a bit less than pleasant. Denticle shape, height, and the number of ridges and cusps differ by species, so sharks can actually be identified based on their skin. Denticle shape can also vary depending on where they occur on the body of the shark.

White shark dermal denticles

When compared to other species, the white shark has very small denticles with thin crowns topped by three ridges that end in rearward-pointing cusps. The roughness of their denticles is similar to 500 grit sandpaper, which is relatively low when compared to other species. Their denticles are also anchored loosely in the skin, which allows for greater flexibility. These characteristics are typical of fast-swimming sharks. The denticles not only provide protection but their design allows water to flow more smoothly over the shark, reducing drag and making swimming more energetically efficient. This reduction in drag enhances thrust and dampens noise so that these sharks tend to move more quietly through the water—this is particularly helpful when moving in for a predatory attack.

TEETH AND JAWS

The business end of any shark is its mouth—there is probably no living maw on the planet that is more feared than that of the white shark. Lined with multiple rows of sharp teeth, the white shark's jaw is an efficient tool for biting and slicing. The teeth are triangular with a single cusp and serrated like a knife, which makes them perfect cutting tools for rending large chunks of flesh from seals and whales. Smaller white sharks, generally less than 9 feet, have teeth that are narrower with one small cusp (or cusplet) on each side. These teeth are well adapted for grasping fish and

The mouth of the white shark is one of the most feared.

squid, which makes up the diet of these smaller white sharks.

The tooth of the juvenile white shark is more narrow.

White sharks have forty-four to fifty-two rows of teeth, generally twenty-five or twenty-six in the upper jaw and twenty-three or twenty-four in the lower jaw. Both the upper jaw, called the *palatoquadrate*, and the lower jaw, called *Meckel's cartilage*, have two parts, a right and a left side separated by a gap called the *symphysis*. When looking directly at the open mouth of a white shark, each side is identical, but tooth shape and size change as you move to the back of the jaw. Like our teeth, the different tooth shapes likely reflect different roles. The teeth in the front two rows on each side of the upper jaw are larger than the others but aligned in a way that forms a continuous cutting edge with the rest of the jaw. The teeth get progressively smaller as you move back, but the tooth in the third row of the upper jaw, called the intermediate tooth, has a curved cutting edge and tapers forward. It is thought that this tooth is placed in the strongest part of the jaw to produce the largest wound. While the upper teeth tend to be slightly broader and more triangular than those in the lower jaw, both sets are built to puncture and slice. By shaking its head from side to side during feeding, the white shark can increase the cutting efficiency of its serrated teeth.

Unlike us, white sharks don't have to worry about losing their teeth. In fact, they literally lose thousands of teeth during their lifetime. If you imagine each row of teeth as a conveyor belt, white sharks never stop growing teeth and always have a sharp set ready to go as new fresh teeth regularly replace the older dull ones. It has been proposed that the frequency of tooth replacement in sharks is associated with tooth wear and not on whether they are broken.

Rows of great white teeth

The jaws of the white shark are suspended by strong muscles and tendons and not attached to the shark's skull. Using these muscles, the white shark can extend or protrude its jaw and teeth to more effectively grab, hold, and bite—I talk about this in more detail in Chapter 4.

A cartilaginous skeleton makes the white shark very flexible.

The white shark can protrude its jaw.

SKELETON

By definition, all sharks have a cartilaginous skeleton. We have cartilage parts as well—just touch the tip of your nose and you will get a sense of what cartilage feels like. It is pretty light and pliable. Therefore, a skeleton of cartilage gives the shark great lightweight flexibility as well as strength and support. Again, for an animal that swims continuously through a dense medium, such a design increases energetic efficiency.

The white shark skeleton is fairly straightforward. It consists of two major parts: the axial skeleton and the appendicular skeleton. The axial skeleton includes the jaws, skull, and vertebral column. As mentioned earlier, the jaws articulate with, but are not attached to, the skull of the shark, which is called the *chondrocranium*, literally "cartilage skull." The

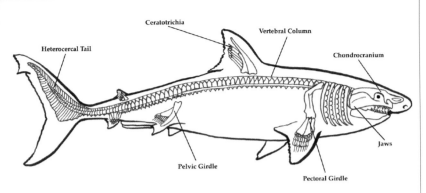

Ceratotrichia

Vertebral Column

Heterocercal Tail

Chondrocranium

Jaws

Pelvic Girdle

Pectoral Girdle

The skeleton of the white shark

chondrocranium houses the brain and sensory centers of the shark and articulates with five branchial arches, which support the gills and all associated blood vessels. The skull is connected to the vertebral column that runs the length of the shark to the tip of the tail. The vertebral column is segmented into 170 to 187 vertebral *centra* (plural for centrum). The spinal cord passes through the neural arch on the top of the vertebral column. Starting roughly around the pelvic fins, two major blood vessels pass through the hemal arch on the underside of the column. The appendicular skeleton includes the girdles and supportive structures of the paired pectoral and pelvic fins as well as the supportive elements (ceratotrichia) of the unpaired dorsal, second dorsal, and anal fins.

For anyone who has seen or held a preserved shark jaw, you can attest to the fact that it looks and feels like bone. Why? Because the cartilage of sharks is covered with a thin, highly mineralized calcium phosphate layer of tiles called *tesserae*, which give shark cartilage the strength of bone without the weight. The white shark and a handful of other shark species were found to have more layers of tesserae surrounding the jaw cartilages than most shark species. This is thought to further strengthen the jaws for biting.

PHYSIOLOGY

The basic morphology of a white shark is interesting, but how it functions, its physiology, is even more intriguing. Telling you about muscle is pretty boring but describing how it works will keep your attention.

SWIMMING

As highlighted above, white sharks have some unique adaptations for keeping the energetic costs of swimming pretty low, including their fusiform body shape, a highly flexible lightweight cartilaginous skeleton, and drag-reducing skin. However, before a white shark can move a single inch, it needs to flex its muscles. Over the course of my career, I've had to catch and handle sharks of all sizes to study them, and I'm always amazed by their muscle mass. Nothing is more exciting, exhausting, or dangerous than trying to restrain an animal armed with sharp teeth and a body composed of over 60 percent muscle. Add to this their flexibility and you might as well be grabbing a tiger by the tail.

Like all fishes, white sharks have two types of swimming muscles: red and white. If you go to the fish market and see fish fillets, you can see the two kinds of muscle. The red muscle is typically only a fraction (less than 5 percent) of the muscle mass and is used for steady swimming at slow speeds. This muscle is powered by oxygen consumption (it is aerobic), so it is loaded with blood vessels for oxygen delivery—hence the color. The white muscle mass, sometimes called fast-twitch muscle is, by far, the greatest proportion (greater than 95 percent) of the total muscle mass. This muscle is designed to contract rapidly with great force for bursts of speed, like those used to elude predators or capture prey. A white shark

exploding through the surface of the water during an attack on a seal is using its white muscle mass. But white muscle is powered anaerobically, meaning that it is not oxygenated and does not have many blood vessels. Anaerobic muscle activity generates lactic acid and a considerable oxygen debt. When this happens to us during periods of extreme exercise, we call this muscle burn, and we are usually physically exhausted. The same is true for white sharks and other fishes, so periods of burst speed are usually pretty short and white sharks use them sparingly. Although we would like to think that white sharks swim fast all the time, racing around the ocean, this is not the case and most of the time they move steadily and slowly, generally between 1 to 3 miles per hour. However,

A white shark explodes through the water's surface in pursuit of prey.

when they do use their white muscle during feeding, their speed increases at least six-fold. The maximum speed of a white shark breaching off the coast of South Africa in hot pursuit of a seal has been estimated at 15 to 22 miles per hour. Not bad for a fish this size.

White sharks are animals, so they need oxygen to live. Like all fishes, white sharks have gills to extract oxygen from the water. Like most sharks, white sharks have five gill slits on each side of their head. In contrast to most bony fishes, which can pump water over their gills, white sharks, and many other shark species, need to swim forward to breath. This is called *obligate ram ventilation*, meaning they're obligated to ram water over their gills to ventilate. During this process, the forward motion of the shark forces water into the mouth, over the gills, and out the gill slits. As water passes over the gills, gas exchange of oxygen and carbon dioxide occurs across the very thin gill filaments, through which blood is flowing. The heart pumps oxygenated blood through the circulatory system to the rest of the shark, where it is utilized to produce energy in a series of chemical reactions called respiration. If a white shark stops its forward movement, it will actually drown, which sounds pretty weird

White sharks have five large gill slits on each side of their head.

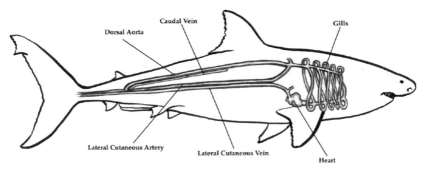

Major blood vessels of the white shark

for such a powerful fish. Since water contains much less oxygen than air, white sharks must breathe ten to thirty times more water to get the same amount of oxygen as a land animal would get from air.

To enhance oxygen delivery to the tissues, large fast-swimming sharks, like the white shark and its closest relatives, have more gill surface area, thicker hearts, and blood that holds more oxygen (more *carrying capacity*) than less active sharks, like the leopard shark. To work more efficiently, the red muscle mass of these active sharks is positioned more centrally and forward in the body than in other species, and it has higher capillary density and myoglobin concentrations for greater oxygen exchange.

ENDOTHERMY

Whenever I go swimming in the ocean, my body loses heat and I get cold. The colder the water, the faster my body cools. Imagine living in the ocean. With very few exceptions, all fishes are *ectothermic*, which basically means cold-blooded.

Ectothermic fish are unable to raise their body temperature more than a degree or two above the water in which they are swimming. Most sharks are no exception, and as a result they need to stay in water of a specific temperature range. The ectothermic blue shark, for example, migrates seasonally to higher latitudes in the late spring and summer as water temperatures rise, but they must also move back as these same waters cool down in the fall. We also need to think of the ocean as a three-dimensional environment, so that same blue shark might dive to over 2,000 feet where the water temperature is less than 50°F but its body cools and it must return to the surface before its body shuts down.

Of all the fishes, more than 34,000 species, there are only six major groups that have evolved *endothermy*, that is, the ability to elevate the temperature of all or parts of their bodies above that of the surrounding seawater. The scombroid fishes, which include the tunas, mackerels, and billfishes, have varying levels of endothermy ranging from the capacity to warm their eyes and brains to elevated muscle and visceral (organ) temperatures. The opah, *Lampris guttatus*, keeps its entire body a bit warmer than the surrounding water. The devil and manta rays in the family Mobulidae appear to have the ability to warm their brains, but their heat production strategy remains a mystery. Only six species of sharks are currently known to be endothermic: the common thresher shark (*Alopias vulpinus* in the family Alopiidae) and the sharks in the family Lamnidae, which includes the great white, shortfin and longfin makos, Pacific salmon, and porbeagle sharks.

The mechanism that keeps these fishes warm is quite amazing. Like all living things, fish create heat when they burn energy—this is a product of metabolism. So, when fish swim, for example, they create metabolic heat. In ectothermic fishes, this heat warms the blood but is quickly lost when the blood reaches the gills, which are in contact with the surrounding (cooler) seawater. Therefore, although all fishes generate heat, very few are able to retain it. The white shark and other endothermic fishes have a complex of specialized blood vessels called the *rete mirabile* or "wonderful net," which is a mesh network of tiny veins and arteries that run side by side. Lamnid sharks have retia mirabilia (plural) close to their red muscle. As warm venous blood

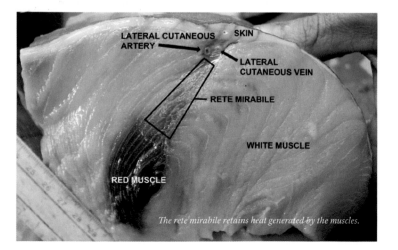

LATERAL CUTANEOUS ARTERY

SKIN

LATERAL CUTANEOUS VEIN

RETE MIRABILE

WHITE MUSCLE

RED MUSCLE

The rete mirabile retains heat generated by the muscles.

leaves the muscle, it flows through the rete where the heat transfers to the incoming arteries and goes back into the muscle. This countercurrent heat exchange system raises the body temperature because the heat is not lost at the gills. In the white shark, the maximum elevation of body temperature over ambient water temperature is 9°F, which means that a white shark swimming in 50°F water can have a body temperature of 59°F. This ability to warm its body is about the same as it is for the shortfin mako, but lower than salmon sharks (15°F warmer). Warmer muscles produce more power and greater speed, and a higher body temperature allows these sharks to move over a much broader geographic range as well as throughout the water column. These sharks also have retia mirabilia close to their eyes, brains, and viscera, making these organs up to 26°F warmer for better visual acuity and more efficient digestion. Because of these highly specialized adaptations, these sharks are among the fastest and the most broadly distributed of all the shark species.

White sharks can dive as deep as 3,000 feet.

DIGESTION

The digestive system of the white shark and other elasmobranchs is unique when compared to other fishes. Sharks use both mechanical and biochemical processes to break down their food and absorb nutrients. The process begins in the mouth and pharynx where prey begins to break down during feeding. Food then moves into the esophagus, a relatively short, muscular tube lined with mucus, which helps to transport food to the stomach. Most sharks, including the great white, have a large, sac-like J-shaped stomach. Second only to the liver, the stomach is one of the largest organs in the shark's abdominal cavity. To aid in breaking down food items, the stomach walls are

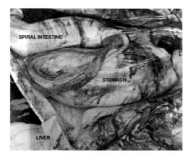

Digestive organs of the white shark

highly muscular with folds called rugae, which allow for the stomach to expand when full. The stomach lining of the white shark secretes very powerful digestive enzymes and acids to break down everything from squid to marine mammal bones. Muscular contractions of the stomach combined with these powerful compounds reduce food to paste for passage to the next digestive organ, the intestines. Sharks need not be concerned with items that they cannot break down; most can quite literally throw up their stomach to be rid of indigestible food items, like turtle shells. This method of rinsing the stomach, then retracting it, is called *stomach eversion.* Although this has never been observed in the white shark, it has in its close relative, the shortfin mako.

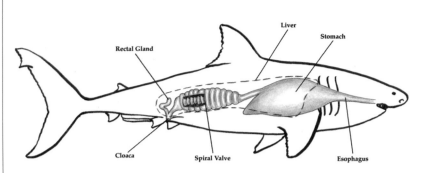

Digestive system of the white shark

From the stomach, the pasty remains of the food pass into an intestine that is unique to sharks and their relatives. The shark intestine contains three sections: the *proximal intestine*, the *spiral intestine*, and the *distal intestine*. Food passing from the stomach through a pyloric valve enters the proximal intestine, where digestive enzymes and bile are secreted through ducts from the pancreas and the liver, respectively. The spiral intestine, also called the *spiral valve*, makes up most of the shark's intestine. In sharp contrast to our small and large intestines, the spiral intestine of a shark is relatively short and stout, only about a foot long in a six-foot white shark. However, the spiral intestine is not a simple tube, but rather a multi-layered organ for maximal absorption of nutrients. To date, four different kinds of spiral intestines have been described in sharks based on their internal structure. For example, one type found in the requiem sharks, looks like a rolled-up scroll. In the white shark and other lamnid species, the spiral intestine is column-shaped and looks like a big corkscrew inside. The shape of the spiral intestine is designed to slow the passage of food. When combined with its huge amount of surface area, nutrient absorption is maximized. After slow passage through the spiral valve, solid matter that is not absorbed or usable passes into the distal intestine, which has a thicker muscular wall for holding waste. Ultimately, this waste is released through the cloaca.

Just how long it takes a white shark to digest a meal remains unknown. As an endothermic species, the white shark maintains an elevated stomach temperature relative to the surrounding seawater. Since the rate of digestion increases with body temperature, digestion is likely faster in the white shark. Although this has never been measured in this species, the digestion rate in the shortfin mako shark has been observed to be less than six hours.

BUOYANCY

Most species of fish have a specialized organ called the swim bladder, which allows them to adjust their buoyancy in the water column. By adding or removing air from the swim bladder, these fish can remain motionless and hover as needed. Sharks and all elasmobranchs lack a swim bladder and must swim forward or they will sink to the bottom. To compensate for the lack of buoyancy control, sharks have a massive liver rich in oils that provide lift. Shark livers, composed of two major lobes, can weigh as much as 30 percent of the total weight of the shark, depending on the species. The liver of the white shark can weigh hundreds of pounds and as much as 28 percent of its body weight.

Up to 90 percent of the liver weight is oil and about 90 percent of that oil is squalene, which is significantly less dense than seawater. This, in turn, reduces the density of the shark by over 90 percent. In the white shark, the liver gets proportionally larger as the shark gets larger. For example, I found the liver of an 8-foot, 325-pound white shark weighed 40 pounds, or 12 percent of its body weight. However, the liver of a 15-foot white shark weighed 551 pounds, which is 20 percent of its 2,750-pound body weight. Think about that, a 500-pound liver! This means that larger white sharks need more lift than smaller ones. The amount of oil in the liver of the larger white shark likely provides more than 100 pounds of lift, which is almost enough to make that shark neutrally buoyant in the water column. The lift provided by the liver, therefore, reduces the amount of energy the white shark must expend to maintain its position in the water column.

This large amount of liver oil is also an important source of fuel to power the shark. For example, the liver of a 1,000-pound white shark contains about 100 gallons of oil, which provides about 2 million calories of energy. Researchers have shown that these large livers provide fuel for long-distance migrations in white sharks. In addition, the liver acts as it does in other vertebrate species by storing other metabolic fuels and vitamins and acting as a detoxifier keeping the shark's general biochemistry well balanced.

In all living things, water always moves from cells with a low concentration of salts or electrolytes (charged molecules) to cells with a high concentration to maintain equilibrium. This is called osmosis. *Osmoregulation* is the process by which animals maintain water balance and the concentration of electrolytes in their bodies from becoming too diluted or too concentrated. Humans generally don't think too much about osmoregulation, but our bodies are doing it all the time. I'm sure there are times when you feel dehydrated and need to drink water. For us, our kidneys are responsible for osmoregulation.

Imagine living surrounded by freshwater or saltwater. Due to osmosis, fish living in fresh water are constantly subjected to an influx of water because their cells are more saline (have more salts) than their environment. On the other hand, marine fish, including sharks, are always threatened by the loss of water from their cells because there is more salt in their environment. To deal with this constant challenge, freshwater fish drink little water and produce large quantities of dilute urine, while most marine fish drink large quantities of water and eliminate salts in small amounts of highly concentrated urine and feces, as well as at the gills.

Living in saltwater, white sharks must osmoregulate to avoid dehydration.

Of course, as we have seen in other aspects of their physiology, white sharks and other elasmobranchs are exceptions. They do not drink seawater. Instead, they retain high concentrations of urea in their tissues and blood to offset the loss of water. Urea is a nitrogen compound that comes from the breakdown of proteins. Most animals, including us, get rid of urea in their urine. To keep urea in their bodies, sharks use their gills, kidneys, and a specialized organ called the rectal gland, which is connected to the lower part of the digestive tract. While the rectal gland removes excess salt, the kidneys retain urea, and the gills help to maintain ionic (electrolyte) balance. The high amounts of urea in sharks allows them to maintain an equilibrium with the surrounding seawater and not dehydrate.

REPRODUCTIVE ORGANS

All elasmobranchs reproduce through internal fertilization, that is, the male mates with the female to transfer sperm for egg fertilization. To do so, the male has two modified extensions to the pelvic fins called claspers, which rotate and insert, usually one at a time, into the female during mating. Without arms and legs, male sharks must rely on their teeth to hold on to females during mating; this results in mating wounds and scars. For this reason, in some species of sharks, the females develop thicker skin to provide protection from the male's sharp teeth.

As you will see in Chapter 4, we don't know a lot about white shark reproductive behavior, but through dissections, we know a bit about their reproductive anatomy. Like most

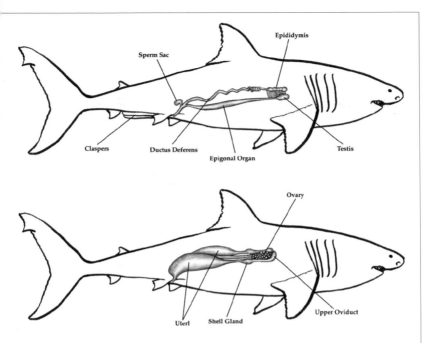

Reproductive organs of the male (top) and female (bottom) white shark

sharks, white shark males have two testes internally located in the forward part of the abdominal cavity. Packets of sperm, called *spermatophores*, are produced by the testes and move down to the claspers through the genital ducts. Sperm is hydraulically pumped through the clasper and into the female by the male's *siphon sacs*, which are located just under the skin forward of the claspers. The female's reproductive system is comprised of a single *ovary* located in the forward part of the abdominal cavity, and paired *oviducts*, *shell* (also called *oviducal*) *glands*, and *uteri*. Eggs produced by the ovary are transported down the oviducts, through each shell gland where the egg is enclosed by a membrane or shell, and down to each uterus. The shell gland is also a site for sperm storage and fertilization for some shark species, but this has not been demonstrated in the white shark.

White sharks are *viviparous*, which means that the fertilized eggs develop in the paired uteri as opposed to *oviparous* or egg-laying species. In viviparous species, the female gives birth to fully developed young,

Paired uteri from a pregnant shortfin mako, the white shark's cousin.

or pups, after periods of several months to two years. In the case of the white shark, the gestation period is unknown, but thought to be about eighteen months. While some viviparous sharks are placental, meaning they have a placental connection between the developing embryo and the mother, the white shark is aplacental with no direct connection. Instead, white shark embryos feed while developing in the uteri. The mother produces additional eggs, which are consumed by the embryos to augment their yolk sac nutrition; this is called *oophagy*. This is characteristic of all species in the order Lamniformes. The sand tiger shark takes this to the extreme and consumes other embryos (intrauterine cannibalism or *embryophagy*), resulting in the birth of a single large (3 feet) pup from each uterus.

There is no indication that white sharks exhibit intrauterine cannibalism. On average, females give birth to five to ten pups in the size range of 4 to 5 feet long.

SENSORY SYSTEMS

All sharks possess an incredible array of sensory capabilities that have allowed them to successfully survive over hundreds of millions of years and exploit diverse habitats around the world. With at least six senses, sharks constantly process incoming sensory information on light, sound, electricity, scent, water movement, and taste. To do so, they generally have large brains, which are comparable in relative size to birds and mammals and largely committed to receiving and responding to sensory information.

The white shark is a top predator that feeds throughout the water column from the surface to the bottom. To do so, it has a well-developed array of senses that are used to locate prey. It is thought that white sharks use different senses depending on their distance from the prey. Imagine a white shark trying to hunt seals while migrating along a coastline. The first sensory cue it is likely to detect is the scent of seals in the water column, so the shark relies on its nose to follow the scent

The white shark has highly developed senses.

trail to its source. As it moves closer, the shark's ear might detect sounds generated by the seals swimming in shallow water. Closing in, the white shark hugs the bottom scanning the surface with its eyes looking for the silhouette of the seal. Once it sees the seal, the shark will initiate a predatory attack, closing the distance rapidly toward the prey. When it is within a few feet, the white shark rolls its eyes back to protect them and relies on electroreception and touch to stay on course and make contact with the seal. With the prey in its mouth, the shark uses taste to confirm that it is indeed a seal. Let's take a closer look at each one of these senses.

SMELL

The sense of smell, or olfaction, has been characterized as the most powerful sense in sharks to the extent that they have been called "swimming noses." Indeed, olfaction is essential for the detection of prey as well as predators and potential mates. Unlike our noses, the olfactory system of fish isn't attached to the respiratory system but remains isolated from the mouth and gills. All fish, including sharks, have external nostrils, called *nares*, which provide access of the surrounding seawater to and from the olfactory organs. Shark nares (singular *naris*), located at the tip of the snout in front of the mouth, are divided by a

The naris of a white shark

skin-covered flap into two areas for incoming and exiting water. The water flows through the nares and into a series of plates containing olfactory receptors, where odors are perceived and communicated to the olfactory bulb of the brain. It is thought that the relative size of the olfactory bulb is related to the importance of olfaction for any given species. The white shark has one of the largest olfactory bulbs measured to date (18 percent of the brain weight). So, in addition to vision, olfaction is more important when trying to find prey than other sensory systems, such as the lateral line and electroreception. It has been suggested that the acute sense of smell in white sharks also aids when navigating over great distances. Much has been said about the ability of a shark to detect the smallest amount of an odor in the water column. Although analogies like "a drop of blood in a swimming pool" do characterize the sensitivity of the white shark's nose, the extent to which sharks respond to these small concentrations remains poorly understood.

SOUND

Water is a much more efficient conductor of sound than air, so sound carries much farther and faster. Sharks do not make sounds, but their environment is loaded with sound generated by both living and non-living things. Although sharks don't have external ears, they have an inner ear that is involved in both sound perception and balance. The auditory (listening) component of the inner ear consists of the *utriculus*, *sacculus* and the *lagena*, which house the sensory components of hearing, the *otoconia* and the *macula neglecta*. This inner ear structure allows for the directional detection of low-frequency sounds up to 1.5 kHz, like those generated by a wounded fish, from distances in excess of 200 yards. For comparison, humans can detect sound over a much broader and higher frequency range of 20 Hz up to 20 kHz.

SIGHT

Situated on the side of the head, the shark's eyes allow for a very wide field of vision with some binocular overlap directly ahead. While the general anatomy of the shark's eye is similar to that of most fishes, there are some unique adaptations including a dynamic iris, large lenses with ultraviolet filters, and a reflective layer called the *tapetum lucidum*, which reflects light back into the retina for increased sensitivity. Although small relative to its body mass, the eyes of white sharks are very large when compared to other species, so vision is thought to play a major role in its feeding ecology (see Chapter 4). As is typical of all eyes, sharks possess an *iris*, the ring-shaped membrane behind the cornea of the eye, with an adjustable circular opening, the *pupil*, in the center. In general, the iris and pupils in sharks differ in color, orientation, and shape between species. White sharks, and other lamnids, have a blue

Eye of the white shark

iris and a horizontal pupil when constricted. This differs from other species whose pupils constrict diagonally or vertically. It is thought that a shark with a horizontal slit is able to resolve detail in the horizontal plane better than in the vertical plane. This might afford the white shark an advantage when hunting prey, but more research needs to be done to confirm this.

Like most fishes, the retina of sharks has both rods and cones as photoreceptors. The rods operate in low light conditions, while cones operate in bright conditions. The rod to cone ratio differs by species and gives us a sense of the light conditions to which that species is adapted. The white shark has four rods to every cone, which is a low ratio compared to other species. This indicates that its eyes are well-adapted to bright light conditions, which we would expect for a highly visual predator that hunts during the day. It also has a high density of photoreceptors in the part of the retina called the *area centralis*. Using its large eye muscles, the white shark can move its eyes to align this part of the retina so it can focus on a target in front of it with binocular vision. This assists the shark in attacking its prey, which it strikes rapidly from below.

Visual information travels from the eyes of the shark to a part of the brain called the *optic tectum*. The relative size of the optic tectum in the white shark is about 40 percent of the total brain, which means that a lot of the brain is dedicated to vision when compared to other senses. However, like all other shark species studied to date, white sharks are color-blind. Although the carcharhiniform sharks have special protective eyelids, called *nictitating membranes,* that extend from the bottom of each eye to protect it when feeding, white sharks do not. Instead, white sharks roll their eyes backward to protect them when striking prey.

The white shark rolls its eye back to protect it when striking prey.

ELECTRORECEPTION

The ability of sharks, skates, and rays to detect electrical fields is a unique sensory capability. To do so, these fish have sophisticated electrosensory organs called the *ampullae of Lorenzini,* which allow them to detect weak low frequency electrical fields. Imagine being able to find your prey by sensing the incredibly low level of electricity that it generates? This would allow you to hunt at night or to detect a critter that is hiding in the sand. The electroreceptors also allow sharks to orient to local magnetic fields, which give them the ability to navigate. Each ampulla of Lorenzini, so named for the scientist that described them in 1678, is a small gel-filled chamber

The small black dots on the nose are the ampullae of Lorenzini.

with sensory cells and a tube-like canal to the surface of the skin. On sharks, the ampullae are distributed and clustered all over the top and bottom of their heads. The sensitivity of these organs is amazing as research shows that sharks are able to detect electrical fields as low as one billionth of a volt. Although this low level of voltage is equivalent to a couple of small batteries spread thousands of miles apart, this is a near-field sense, which means that it is used by sharks when within a few feet of prey. In the white shark, the electrosensory system is thought to provide information on the location of the prey during the final moments of the attack, during post-attack manipulation or pursuit of prey, or to detect changes in the disposition of the prey such as bleeding.

Many of the ampullae on the top, bottom, and sides of the head and snout are within the visual field, which is thought to provide an excellent means to detect and track nearby prey when the eyes are rolled back, at night, and when the prey is beneath the snout.

TOUCH

All fish have a specialized sensory system called the *lateral line*, which allows them to detect water movements; this is called *mechanoreception*. Sensory receptors lying along the surface of the shark's body in low pits or grooves detect water displacement and, therefore, give the shark the sensation of feeling its surroundings, a bit like our feeling of touch. This allows the shark to orient to water currents, avoid predators, detect prey, maneuver

Lateral line of the white shark

around obstacles, move in schools, and physically interact with prey and other sharks. The mechanoreceptors are called *neuromasts*, which are composed of a group of sensory hair cells covered with a gelatinous cupula. On sharks, neuromasts can be found in several different arrangements. Those in the lateral line canals run along the length of the body to the tip of the tail as well as traversing various parts of the head.

Individual neuromasts, called pit organs, are positioned all over the body of the shark, on the tail, and between the pectoral fins. This configuration can differ between species. For example, the spiny dogfish has less than eighty pit organs per side while the scalloped hammerhead has more than 600. Unfortunately, the pit organs have not been well-studied in the white shark. A fruitful area for future research!

TASTE

Gustation is the sense of taste and, like olfaction, it is thought to play an important role in feeding. Taste is perceived by taste bud receptors on *oral papillae* that have been mapped in the mouths and gills of some sharks, but not in the white shark. These oral papillae, which develop early in shark embryos, allow sharks to evaluate the suitability of potential prey, leading to ingestion or rejection of the item. For sharks, it has been suggested that papillae size and density are related to taste sensitivity—the bigger and denser the papillae, the greater the capacity to taste. Although the oral papillae have not been studied in the white shark, they are relatively small in the closely related shortfin mako, which suggests that this species relies less on gustation during feeding than other species. The mako is a predator that relies more on vision and olfaction than gustatory discrimination. This might be the case for white sharks, as well. Unfortunately, studies on taste in sharks are lacking and there is clearly a need for more research on the sensitivity of their taste buds, behavioral preferences to specific chemicals, and the role of taste buds in food choice.

3
DISTRIBUTION AND MOVEMENTS

Sharks are difficult to study. Think about it. They live in the ocean, which is a place very different from where we live. The ocean is inhospitable to us, so studying anything that lives there is not easy. Imagine an 18-foot female white shark swimming 1,000 miles off the east coast of the United States in deep blue water. Where did she come from and where is she going? Is she going anywhere at all? How does she spend her day, swimming at the surface or on the bottom? How does she interact with the environment? These are all questions related to the distribution and movement ecology of this species.

When we discuss the distribution of any species, including the white shark, we really need to take into consideration the movements of that species. That is because distribution can change over multiple geographic scales depending on time of year, the size of the shark, and life stage. Movements can range from very fine-scale behavior over minutes, hours, and days to broad-scale migration over weeks, months, and seasons. Also, when it comes to sharks and other marine animals, we must remember that they live in a three-dimensional environment. So, we can describe their distribution and movements as both horizontal and vertical. For example, a white shark might have a horizontal location off Massachusetts as well as a vertical location at 50 feet deep. The next step for scientists is to relate the three-dimensional distribution of a species to its environment. This is called movement ecology, which is now being studied extensively for many shark species, including the white shark. How do we do that?

As is the case for most shark species, our historical knowledge about white shark distribution comes from those who spend their time and, in most cases, earn a living on the ocean, namely fishermen. For as long as humans have fished, they have encountered white sharks. Over the centuries, fishermen's logs have helped us to map the distribution of this species and to study some aspects of its movement ecology. Even today, by knowing where and when fishermen catch sharks, we can get a sense of how shark populations are structured

and whether they are changing. Data obtained from fishermen is typically referred to as "fisheries-dependent" because we depend on fishermen to provide it. While this is a great source of information, these data can also be biased by when, where, and how fishermen are fishing. For example, if fishermen don't report white sharks from a particular region of the ocean, does that mean they are not there? Maybe, but it could also reflect the fact that the fishermen are not fishing that area! Fisheries-dependent methods, like catch records and conventional tagging, are a great source of information, but we also need fisheries-independent sources.

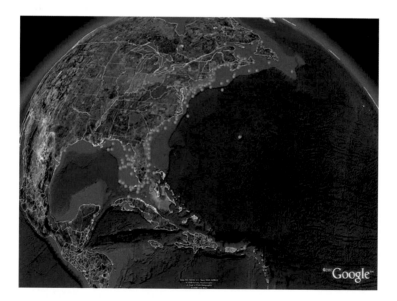

The distribution of the white shark in the western North Atlantic based on fishermen's reports.

Over the last two decades, the technology used to study the distribution of marine animals, like white sharks, has exploded. Smaller and better batteries, more sophisticated sensors, and satellites have led to the development of new tagging technologies that are independent of fishermen's reports. These tags allow us to examine shark movements and distribution regardless of where and how deep the animal swims. In recent years, these fisheries-independent methods have really opened our eyes to the extensive distribution, migrations, and behavior of the white shark.

For more than half a century, scientists have used a variety of tags to study the movement ecology of white sharks. In the following sections, I briefly review each of these methods, ranging from the simplest to the most sophisticated. Each tag type has its advantages as well as disadvantages, and the choice of which to use has a lot to do with the scale of the study (fine-scale behavior vs. broad-scale migration), the duration of the study, and, ultimately, the objectives of the study.

The historical method used to investigate shark movements is conventional tagging. The National Marine Fisheries Service (NMFS) Cooperative Shark Tagging Program in Rhode Island is the largest and longest running shark tagging program in the world. Started by Jack Casey in 1962, this program has enlisted the service of more than 8,000 cooperating recreational and commercial fishermen, fisheries observers, and biologists tagging sharks using conventional tags. Through their efforts, more than 250,000 sharks of thirty-five species have been tagged and 14,000 sharks of thirty-one species have been recaptured.

When a shark is tagged, a small, individually numbered marker is placed on the fish, either through the dorsal fin or held in place by a small dart inserted into the muscle, and it is released. When the shark is tagged, the species, date, location, size, and sex are recorded and entered into a database. If the shark is recaptured by a fisherman and the same data are reported, we can use those two data points to examine distribution and movement from the tagging location to the recapture location. By far, most (51 percent) of the tagged sharks are blue sharks, which have been recaptured all over the North Atlantic. By tagging and recapturing hundreds of blue sharks, a pattern of migration emerges

A blue shark with a conventional tag at the base of its dorsal fin.

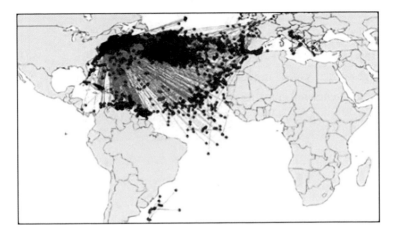

Movements of the blue shark in the North Atlantic based on conventional tags.

for that species. Tagging data also yield rough estimates of swimming speed as well as growth rates. On average, about 6 percent of tagged sharks are recaptured, so this method does require a lot of tagging.

Ultimately, tag-recapture data yield information on the horizontal movements of sharks from point A to point B, but no sense of what the shark did vertically in the water column. Conventional tagging is also fisheries-dependent because it relies heavily on the cooperation and fishing effort of fishermen, so the results can be biased. Unfortunately, the white shark is historically an elusive species and less than sixty have been tagged with conventional tags in the Atlantic. Given the recapture of only two of these fish, conventional tagging has not proven to be a viable tool to study the movement ecology of this species.

The use of electronic tags to study shark movement dates back to the 1960s. Acoustic tags are transmitters that emit high frequency pings that can be followed by the researcher or picked up by an array of receivers. There are two kinds of acoustic telemetry: active acoustic telemetry produces short-term datasets; passive acoustic telemetry produces long-term datasets.

Active acoustic telemetry can be used to examine the fine-scale, three-dimensional movements of a shark for up to several days. The first researcher to actively track a white shark with this technology was Frank Carey from the Woods Hole Oceanographic Institution back in 1979. He placed an acoustic transmitter on a white shark that was feeding on a dead whale off Long Island, NY, and followed it for 3.5 days. The high frequency sound pulses emitted by the transmitter were picked up by a hydrophone mounted to the boat and relayed to a receiver, which was monitored by the crew. By pointing the vessel in the direction of the sound, they were able to follow the white shark. Using the position of the boat as well as temperature and pressure data sent by the tag, Frank and his team were able to examine the fine-scale behavior of the white shark for the very first time. After the tracking was over, they not only had information about where the shark moved and its swimming speed, but also the depth and temperature of the shark. They learned that the white shark moves on average about 2 miles per hour and swims up and down from

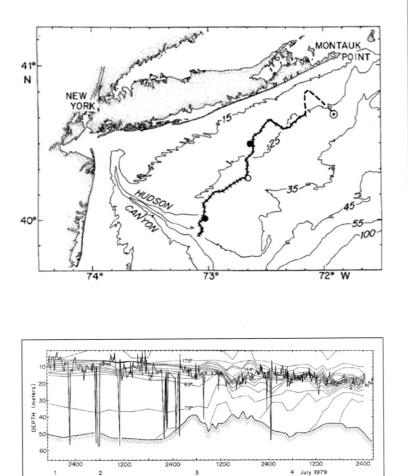

Horizontal (top) and vertical (bottom) movements of the first white shark tagged with an acoustic transmitter in 1979.

the surface to the bottom, but spends most of its time in the thermocline at about 30 to 50 feet deep where water temperature changes dramatically from the warm mixed surface waters of 64°F to the cooler deeper water of 48°F. They also observed, for the first time, that white sharks are endothermic and able to elevate their body temperature up to 9°F warmer than the surrounding water. Active acoustic telemetry has its limitations: it's expensive, time consuming, and terribly short-term, lasting typically only a couple of days. So, you may get great high-resolution, three-dimensional data, but only over small periods of time.

Passive acoustic telemetry works off the same principle but relies on an array of acoustic receivers or listening stations to track the sharks instead of a boat. These transmitters don't have to ping as frequently

Researcher tags a white shark with an acoustic transmitter.

because no one is actively following the sharks, so they last a lot longer, up to ten years! Researchers all over the world have used this technology to study the seasonality, habitat preferences, and residency of white sharks. For example, off the coast of Massachusetts, we have tagged over 300 white sharks with these acoustic transmitters. To track their movements, we've set an array of more than 100 receivers all along the coastline. When one of these sharks enters the range of one of these receivers, which is typically about 100 to 500 yards depending on conditions, the receiver logs the time, date, and identifying number for that individual shark. We can then offload those data and examine when that shark arrived, where it spent its time, and when it left. Using this technology, we have learned that white sharks arrive in Massachusetts waters as early as mid-May and leave as late as mid-December, but peak season is August through October. They prefer to spend their time along the eastern shoreline of Cape Cod where there is the highest abundance of seals.

A white shark with an acoustic tag.

While passive telemetry allows for long-term studies on individual sharks, this only works if the shark remains within the range of the listening array. Fortunately, a lot of researchers use this technology on a variety of species all over the world. These receivers can detect all animals tagged with this technology, including white sharks. So, for example, our tagged white sharks have been detected by other researchers from the Gulf of Mexico, all along the eastern seaboard of

Acoustic receiver that detects tagged white sharks.

the United States to as far north as Newfoundland, Canada. This allows us to examine broad-scale movements as well. While passive acoustic telemetry is a great fisheries-independent approach to examining white shark movements, there are very few receivers deployed farther offshore, so we are only monitoring coastal areas and, therefore, missing the open ocean. For this reason, we need to also use satellite-based technology.

SATELLITE-LINKED TAGGING

One of the more recent innovations in tagging technology combines all of the positive aspects of conventional tagging with those of active acoustic telemetry. That is, a long-term fisheries-independent tag that produces movement data without having to follow the shark or retrieve the tag. These tags

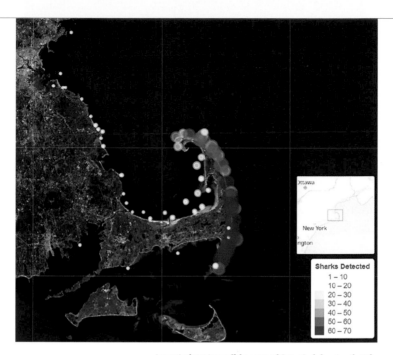

Acoustic detections off the coast of Cape Cod showing that the greatest number of white sharks are off the eastern shoreline.

contain transmitters that communicate directly with an array of satellites orbiting the planet. There are basically two general kinds of satellite tags, the real-time positioning tag and the pop-up archival tag. Both have produced new and amazing information about white shark distribution and migrations throughout the world. Without this technology, we never would have learned that white sharks spend a lot of their time in the open ocean diving to incredible depths.

Any time a shark with a real-time positioning tag, sometimes called a Smart Positioning and Temperature Transmitting (SPOT) tag, comes to the surface, it transmits to a satellite, which calculates and emails its position back to the researcher. So, we can get almost real-time updates on where the shark is going horizontally. That information can be posted to a website or app, like Sharktivity, so the public can also see what the sharks are doing. These tags can also produce multiple years of data. While this technology sounds ideal, there are limitations. First, the tag must be as high as possible on the shark so it can transmit when it breaks the surface. That means the shark must be captured and the tag bolted high on the dorsal fin, which is the only part of the shark that generally comes out of the water. It has been shown that this process

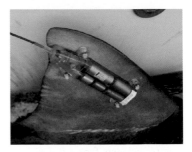

Real-Time Positioning Tag

not only stresses the shark but can also cause damage to the fin over time. Second, sharks don't breathe air like sea turtles and whales, so they don't need to come to the surface. There are no guarantees that the researcher will ever hear from a tag because many sharks stay associated with the bottom. No data, no good. Third, the tags don't give us any vertical data for the shark—the horizontal data are only two-dimensional with no depth information.

Because of these limitations, many white shark researchers prefer to use pop-up satellite archival transmitting (PSAT) tags. These tags are placed on the shark like conventional and acoustic tags, but they contain a microcomputer that allows the researcher to program the tag to release from the shark. While on the shark, the tag collects and archives water temperature, depth, and light level data. After release, the tag floats to the surface and transmits the archived data to a satellite, which relays them back to the researcher. With these data, you can see where the shark traveled both horizontally and vertically as well as the shark's preferred water temperatures. These tags are more expensive, but they can be deployed for up to a year on free-swimming sharks (no stress) and they generate a lot of three-dimensional data.

Collectively, the real-time and PSAT tags have produced a remarkable amount of information about the movements of white sharks. Until the advent of this technology, it was generally thought that white sharks were a coastal species, preferring to remain associated with land masses. Now we know that this is clearly not the case. These tags have demonstrated that white sharks spend a lot of time in the open ocean diving to incredible depths through a very broad temperature range.

Pop-up satellite tag

GENERAL DISTRIBUTION

If you pick up any fish guide or search the web for a white shark distribution map, you see that it is one of the most broadly distributed shark species on Earth, occurring in the Atlantic, Pacific, and Indian Oceans from the tropics to temperate latitudes and from coastal waters to the open ocean. In contrast to what most people think, the white shark is largely a temperate water species, meaning it prefers the middle latitudes and cooler, non-tropical water temperatures.

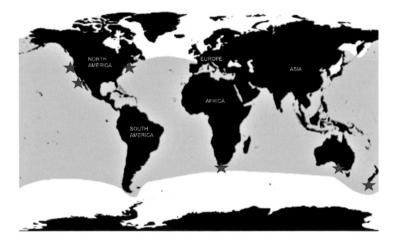

Global distribution of the white shark (grey) with hotspots (stars).

ATLANTIC OCEAN

In the Northwest Atlantic, the white shark is documented from Newfoundland to the Caribbean, including the Gulf of Mexico and Bermuda. It is rare in the eastern North Atlantic, although there are catch records from the Azores, from the west coast of France and Portugal south to the Canary Islands, and along the west coast of Africa to Ghana. The historical presence of the white shark is well-established in the Mediterranean Sea, but no hotspots have been identified to date. In the Southwest Atlantic, the white shark is rare with only a handful of records from Brazil, Uruguay, and Argentina. In the Southeast Atlantic, white sharks have been documented from Angola to South Africa, with the latter having the greatest abundance.

INDIAN OCEAN

In the Western Indian Ocean, white sharks are found from South Africa north along the eastern side of Africa, including Mozambique, Tanzania, Kenya, Madagascar, the Seychelles, and Mauritius. A single specimen has been reported from Sri Lanka. In the Eastern Indian Ocean, the species is well documented in Australia from Western Australia along the southern coast to South Australia, Victoria, and Tasmania on the eastern border of the Indian Ocean.

PACIFIC OCEAN

In the Southwest Pacific Ocean, the white shark occurs along the eastern coastline of Australia from New South Wales to Queensland, New Zealand, and New Caledonia. On the other side of the South Pacific, there are a few records from the Galapagos Islands, Peru, and Chile, but sightings are extremely rare. In the Northwest Pacific, the white shark is well-documented from Russia south to Korea, Japan, China, Taiwan, the Philippines, and Vietnam.

In the Northeast Pacific, the white shark is broadly distributed from Alaska and Canada, south along the entire United States Pacific coast to Mexico.

AGGREGATIONS

The white shark is a unique species of shark in that it migrates to very specific locations at certain times of the year to feed on *pinnipeds*, which is the group of animals including seals and sea lions. These white shark "hotspots" are places where white sharks aggregate to feed and perhaps mate, although mating has never been observed. Almost everything we know about white shark movements and behavior has come from observing and tagging animals at these various aggregation sites. Because these hotspots are predictable locations to find white sharks, they also attract filmmakers and eco-tour operators (see Chapter 6)—people who want to showcase and dive with these animals.

There are currently six white shark aggregation areas in the world, located

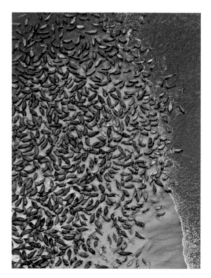

Seal and sea lion colonies create hotspots for white sharks.

off the coasts of Central California, Mexico, Massachusetts, South Africa, South Australia, and New Zealand. Each of these areas is remarkably different both above and below the surface. They contain different species of pinnipeds, water depths ranging from deep to shallow, and visibility from clear to coffee. The ability of the white shark to successfully survive in each of these areas really highlights the adaptability of this species.

CENTRAL CALIFORNIA AND GUADALUPE ISLAND, MEXICO

These two hotspots are off California and Guadalupe Island in the Northeast Pacific. They are among the most well-studied white shark locations in the world. Since the late 1960s, researchers have been observing and studying white sharks off Central California at the Farallon Islands. Also referred to as the Devil's Teeth, these islands lie about 27 miles east of San Francisco. White sharks feed on juvenile northern elephant seals and California sea lions during the fall months. Guadalupe Island, Mexico, supports four species of seals and sea lions and white sharks are abundant

Elephant seal

through the summer and fall months. Tagging studies indicate that white sharks tagged in these two areas move in the winter to tropical and subtropical waters in the middle of the Pacific Ocean between Baja California and Hawaii known as the White Shark Café, which is about 320 miles wide. While in this oceanic phase, white sharks dive to depths of more than 1,000 feet. Some sharks move as far west as the Hawaiian Islands. Although white sharks from both aggregations comprise a single Northeast Pacific population, Guadalupe sharks rarely go to Central California, and vice versa.

Guadalupe Island

CAPE COD, MASSACHUSETTS

In the Northwest Atlantic, the white shark is broadly distributed from Newfoundland, Canada, south to the Caribbean, including the Gulf of Mexico and Bermuda. Over the last decade, Cape Cod, Massachusetts, has emerged as an aggregation site for this species due to the restoration of the gray seal population in this region. White sharks aggregate at Cape Cod during the summer and fall months. Tagging data indicate that this species migrates into northern latitudes in the summer months and moves south in the winter. Some large subadult and adult white sharks migrate into the open Atlantic in the winter as far east as the Azores, but these movements appear to be nomadic without a focal area like the White Shark Café in the Pacific. While in the open

Atlantic, white sharks dive to depths in excess of 3,700 feet through water temperatures ranging from 35°F to 87°F.

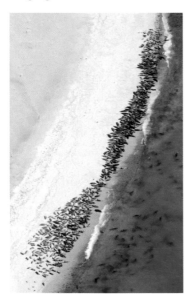

Gray seals on the eastern shoreline of Cape Cod

WESTERN CAPE, SOUTH AFRICA

The province of Western Cape, South Africa, is the southernmost region of this country and the location where the Southeast Atlantic Ocean borders the West Indian Ocean. The white shark is broadly distributed in this region, but the most heavily studied hotspots are False Bay, Gansbaai, and Mossel Bay. Although white sharks can be found year-round throughout the region, they aggregate during the autumn and winter to feed at island rookeries of Cape fur seals. Sharks tagged at these sites make broad-scale migrations of more than 1,200 miles along the South African coast as far north as Mozambique and back again. Some sharks exhibit offshore movements into the Indian Ocean, with one white shark going all the way to Western Australia and back again.

NEPTUNE ISLANDS, AUSTRALIA

In Australia, the white shark is broadly distributed along the southern half of the continent

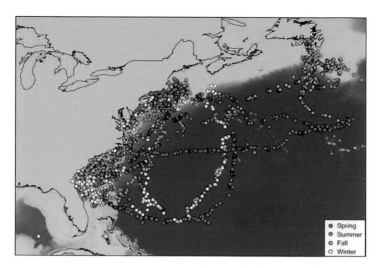

Broad-scale movements of white sharks in the Atlantic.

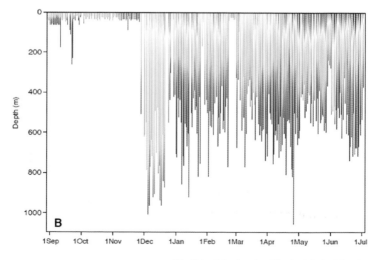

The diving behavior of a white shark in the Atlantic.

from Queensland to Western Australia. White sharks aggregate at the Neptune Islands off South Australia to feed on long-nosed fur seals. While males visit these islands year-round, females are only there in the fall and winter, likely to feed on lactating fur seal mothers and their pups. Tagging data indicate movements northward to tropical and subtropical waters along the east coast of Australia during the late fall and winter months as well as excursions into offshore oceanic waters.

NEW ZEALAND

White sharks are common in New Zealand waters. The two predominant aggregation sites are the Chatham Islands, 400 miles off the east coast, and Stewart Island off the southern tip. Both locations support large breeding colonies of long-nosed fur seals. White sharks aggregate at these sites from the late summer to early winter, peaking in the fall. Tagging data show that these white sharks migrate offshore annually to subtropical and tropical regions of the Southwest Pacific. Unlike white sharks in the Northeast Pacific, these sharks don't go to a specific area like the White Shark Café. Instead, they spread out over a vast area of the ocean, diving to depths as great as 4,000 feet, with some moving as far west as Australia.

A fur seal basks on the shoreline of North Neptune Island

White sharks make long distance migrations into the open ocean

4
LIFE HISTORY

L et's imagine a big female white shark swimming in the open Atlantic. What does she eat and how often? Does she have any enemies? Is she mature? If so, when and where will she mate? How long does it take for her young to be ready for birth? Where and when does she give birth? How old is this fish and how long does she live? How fast does she grow? These very basic questions are all related to the life history of this shark.

How would you answer these questions? The shark is too big to keep in captivity. The shark is too far offshore to directly observe. Even if you could find her, it's unlikely that you would be able to stay with her without influencing her behavior. The bottom line is hard to believe, but after centuries of studying sharks, we still have so few answers to these seemingly simple questions. In this chapter, we look at the life history of the white shark—what we know and what we don't know. In doing so, I give you a sense of how we go about answering these questions.

FEEDING ECOLOGY

Arguably, the most conspicuous anatomical feature in sharks is their set of jaws. It is certainly highlighted by Hollywood producers in almost every film that includes a shark. As much as the jaws can be viewed as the most frightening part of any shark, it is important to remember that this is the body part that sharks need to stay alive. Like our mouths, sharks use their jaws for energy intake. There are many kinds of sharks with many kinds of jaws, and a variety of ways sharks use their jaws. Some species rely on suction feeding, others ram and bite, and some even use their mouths to filter tiny plankton from the water. The white shark is a ram and bite feeder. Like the tiger and requiem sharks, white sharks use the

The white shark is a ram and bite feeder.

subterminal (under the head) placement of the mouth to generate large biting forces with the huge muscle mass supporting the jaws. Coupled with a protrusible jaw that can be extended forward, this great force allows white sharks to cut through the hardest of prey items, including the skin, blubber, bone, shell, and muscle of fish, marine mammals, and turtles.

When we watch footage of a white shark biting something at the surface, it happens incredibly fast. But when we slow down that footage, we see that the shark goes through a series of distinct steps with each bite. As it closes to within three feet of its prey, the shark first lifts its snout and drops its lower jaw, expanding its mouth as wide as possible to a maximum opening or gape. This allows the shark to take the largest bite possible with each attempt. Once fully opened, the shark's upper jaw, also called the palatoquadrate, is then protruded and extended forward out of the mouth, fully exposing its upper teeth as it starts to close. At the same time, the shark moves its lower jaw, called Meckel's cartilage, upward and the mouth closes down on the prey. All of these steps happen in less than one second and as fast as two-tenths of a second.

Ever since I was a kid, I've heard that sharks can exert incredible force when they bite down, but this largely came from people feeding hard objects to sharks and not conducting real experiments. Computerized tomography (CT) scanning, which creates cross-sections of the jaw, and three-dimensional model simulations have now allowed researchers to generate more realistic estimates of bite force in animals that are difficult to keep in captivity, including the white shark. These models show that the bite force of the largest white shark is over 18,000 newtons or 4,000 pounds, which is the highest calculated for any living species. However, when these estimates are adjusted for body size, they are not very high if compared to mammals or reptiles of similar size. For example, a 500-pound African lion can exert a bite force more than twice that of a 500-pound white shark. But the white shark might not need comparable bite force because it has slender, serrated teeth that are extremely effective at penetrating potential prey.

DIET

What do white sharks eat? There are some who believe that white sharks eat people. This is simply not true. If sharks ate people, we would have a lot more shark attacks every year. In fact, with so many people flocking to the coastline for a variety of water-related activities, we would be providing plenty of food for them every day. Instead, white sharks are well known for attacking and consuming seals, sea lions, and other large prey, but they don't exhibit these food habits through-out their lives. After all, it's hard to imagine a newborn white shark being able to successfully kill a seal. Instead, white sharks go through an

Seals are part of the white shark's diet.

ontogenetic diet shift, which means that their diet changes as they get older and larger. This has been demonstrated by researchers using a couple of different scientific methods.

While it would be great fun to swim with white sharks all day and observe what they eat, this is hardly the easiest or the most realistic way to examine their diet. With almost thirty years of diving with white sharks, I've never been underwater and watched one eat unless we were feeding it. However, white sharks do feed on seals and/or sea lions at aggregation sites, like South Africa and Cape Cod, and these areas provide opportunities to directly observe not only what they eat, but how they eat. These observations, though, only provide a snapshot of the white shark diet because not all white sharks go to those locations, and they are not typically there year-round.

To really understand the diet of any species, you need to sample all sizes of that species, from newborn sharks to mature adults, over the course of the entire year. The traditional way to study this is to look in a white shark's stomach. For decades, scientists have been sampling white sharks landed by fishermen to examine the contents of their stomachs.

A researcher dissects a white shark.

However, this obviously involves lethal sampling and now that white sharks are rightfully protected in many parts of the world, access to these samples has diminished. To avoid having to kill the shark, some scientists are taking advantage of the fact that sharks can evert their stomachs. By inserting a tube down a live shark's throat and slowly retracting it, the stomach is pulled out and its contents removed. A similar method called *gastric lavage* involves pumping seawater down a tube inserted in the shark's stomach and flushing the contents out the mouth. Although these methods sound traumatic, they do not cause damage to the shark and are clearly much better than killing it. Of course, these methods do not work well on large sharks, which are difficult to handle. Another more recent, noninvasive technique for understanding what a shark eats involves examining the nitrogen and carbon composition of its tissues. Basically, the premise of the analysis is that you are what you eat. This method, called *stable isotope analysis*, does not give you detailed information on diet, but does reveal where the species fits into the overall food web or, in other words, it's *trophic position*.

When it comes to the white sharks, researchers have used direct observations of predatory behavior at aggregation sites, stomach contents, and stable isotope analysis to determine that the white shark is a generalist, opportunistic predator at the top of the food web, also called an *apex predator*. The stomachs of white sharks have yielded a variety of fishes, invertebrates, marine mammals, and other animals, like turtles and sea birds, as well as some garbage, but no license plates! The fish species include those that live in the water column in large schools, those that live on the bottom, and a variety of other small to medium-sized sharks and rays. The dominant inver-

The stomach contents of a white shark contained this large fish and small seal.

tebrate is squid, but crabs have also been found. Marine mammal species include seals and sea lions (pinnipeds), dolphins and porpoises, and larger whale species, which are scavenged.

Although the white shark feeds on a wide variety of animals over the course of its life, the composition of its diet changes as it gets larger, as does its ability to target larger prey. Small white sharks from the size at birth to about nine feet in length are built for speed and agility. They are relatively slim, like mako sharks, and well-adapted for feeding on schooling fish, squid, and other sharks. The teeth of small white sharks are relatively long and narrow and perfect for grasping prey like small fishes. That's why the diet of these smaller white sharks is dominated by bony fishes and small sharks. As white sharks get larger, their morphology transforms like a teenager

A white shark scavenges a large whale carcass.

going through puberty. This happens when they generally reach eight to nine feet in length but could happen at smaller or larger sizes depending on the individual. Their body shape gets bulkier as their muscle mass increases and their tooth shape becomes broader and more triangular, which is ideal for cutting and gouging pieces from prey too large to swallow whole. Using speed and powerful jaws, these larger white sharks are now capable of attacking and killing animals, including pinnipeds (seals, sea lions) and dolphins and porpoises. They are also capable of tearing large chunks of blubber and muscle from dead whale carcasses, which they opportunistically scavenge.

As a generalist feeder, the white shark's diet also reflects the most dominant species where it lives depending on the time of year. For example, small sharks, like dusky sharks, are a major part of the diet of young

White sharks migrate great distances to feed on sea lions in some parts of the world.

white sharks off South Africa, while bottom fishes, like hake, are found more frequently in small white sharks sampled off the northeastern United States. It has been shown that larger white sharks migrate great distances seasonally to locations where seals, sea lions, and, perhaps, dead whales are more likely to be encountered. As we discussed in Chapter 3, these locations become white shark aggregation sites or hotspots where scientists, ecotours, and filmmakers have predictable access to this species.

One of the big questions often asked about white sharks is: How often do they have to eat? As you can imagine, this is a tough question to answer because it is difficult to observe the feeding frequency of an individual white shark over many days without losing sight of it. Also, there is a big difference between how often the shark theoretically eats and how often it actually eats. Sharks are all about energy efficiency, whether swimming or feeding, and the white shark is no exception. In general, sharks can sustain themselves on small and infrequent meals and their slow digestive rates mean that sharks burn their fuel, oxygen, and food very efficiently. This "burn rate," so to speak, is called *metabolic rate*. Think of it as the fuel efficiency of your car. Some cars burn a lot of gas to get from the house to the grocery store and some don't. Sharks are very efficient cars—their metabolic rates are low when compared to other fishes.

As an opportunistic predator and scavenger, the white shark likely eats as frequently as it can. Nonetheless, scientists have made efforts to estimate how long a robust meal might last before the white shark has to feed again. This was first done by researcher Frank Carey, who acoustically tracked a 15-foot white shark off the coast of New York in 1979 (see Chapter 3). Based on internal temperature changes measured in the shark and sent back to Carey, he estimated the metabolic rate and determined that 60 pounds of whale blubber would sustain the shark for at least 1.5 months. However, this was likely an overestimate. More recently, researchers used estimates of swimming speed from the acoustic tracking of multiple white sharks as well as metabolic rate estimates from captive white sharks and closely related mako sharks to show that 60 pounds of blubber would only last about twelve days, which is still pretty long. Because this work was done in Australia, where white sharks are thought to target long-nosed (or New Zealand) fur seal pups, these authors estimated that a white shark needs to feed on at least one 30-pound fur seal pup every three days. This would certainly explain why white sharks spend a lot of time at seal colonies.

This whale carcass has multiple bites from white sharks.

PREDATORY BEHAVIOR

It's one thing to know what white sharks eat and another to know how they feed. Most of what we know about white shark predatory behavior comes from direct observations at aggregation hotspots. But when we look at the six hotspots around the world, we see significant habitat and prey differences. For example, Cape Cod is characterized by shallow, shoaling turbid coastal waters inhabited by gray seals. In sharp contrast, Guadalupe

Off Cape Cod, MA, white sharks hunt gray seals in shallow water.

and mouse, which involves constant pursuit, repeated escapes, and some captures.

White sharks and seals play a complex game of cat and mouse.

Island has deep, crystal clear waters inhabited by four species of seals and sea lions. As a species, the white shark is the same throughout the world, *Carcharodon carcharias*, but it has adapted a successful predatory strategy in each of these very different locations. This certainly highlights the behavioral plasticity in this species and its ability to survive.

What we've learned is, like any predator-prey relationship, white sharks and seals play a very sophisticated game of cat

After all, the white shark needs to eat, but the seal does not want to be eaten. While the white shark has size, speed, and power, the seal is more agile, has excellent vison, and is likely more intelligent. The white shark is an ambush predator, so feeding success requires speed

Seals are more agile than large white sharks underwater.

and stealth. Once the shark is detected by the seal, the game is over, and the shark loses. For many years, researchers were convinced white sharks used a "bite and spit" strategy. Based on observations of white sharks feeding on bait, reports from shark attack victims, and wounds on seals and sea lions, researchers not only noticed that white sharks roll their eyes back when striking, presumably to protect them, but also that they move away from the injured animal after rendering a big bite to allow the seal to bleed to death. This was thought to eliminate the risk of injury from the seal, which possesses sharp claws and teeth. Once the injured animal

A white shark ambushes a seal off Cape Cod, MA.

dies, it is then consumed by the shark.

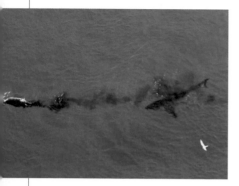

Some white sharks bite and spit out the seal so it bleeds out.

However, more and more observations have been collected of white shark predatory behavior, and the "bite and spit" approach does not appear to be used consistently around the world. By far, the most studied hotspots are Southeast Farallon Island off Central California and Seal Island in False Bay, South Africa. At these locations, white sharks have been routinely and intensively observed preying upon seals and/or sea lions. This has allowed these researchers to show patterns in white shark hunting behavior as they relate to a variety of environmental and other factors. At Southeast Farallon Island, researchers have been observing white sharks feeding on juvenile northern elephant seals and California sea lions since the 1970s. Peak activity occurs during the autumn months.

Seal Island in False Bay, South Africa

The overall predatory strategy is the white shark remains at depth and strikes its prey from below and behind with stealth and speed. To remain undetected, the white shark hunts from a depth where it can't be seen by the seal. Its dark back certainly helps. Much depends on water clarity. These sharks search for prey around the island over days or weeks, traversing the area in a manner that maximizes coverage, and swimming close to the bottom to remain undetected by the seals. They tend to have the greatest success at entry and departure points for seals around the islands when the tide is high during daylight hours. Instead of the standard "bite and spit" strategy, their attack depends on their prey. Northern elephant seals are generally attacked below the surface, then continue to be chased or carried around by the shark underwater until dead. The shark then feeds on the seal at the surface because the carcass floats. However, sea lions are typically attacked at or near the surface in a more explosive strike causing a massive splash of water. In most cases, the injured sea lion usually survives the initial strike

White sharks strike from below with stealth and speed.

and surfaces disoriented, only to be pursued and ultimately killed by the shark. In both cases, the prey might be carried hundreds of yards away before being consumed. This might be to eliminate competition from nearby sharks.

White sharks sometimes carry their prey away before eating it.

In South Africa, water clarity is typically about 20 feet, so the most successful white sharks remain deeper at about 85 to 100 feet. They also hunt primarily around sunrise during a period of low light conditions. At these depths and low light levels, the shark can easily detect the seal silhouetted against the bright sky while the seal cannot see the shark, which blends with the darkness below. The peak season for white shark predation on Cape fur seals at Seal Island is from late June to early August (winter) and the sharks are there in high densities. They are targeting very young Cape fur seals, born the previous November/December, that come and go from Seal Island to forage offshore. Most of the attacks are clustered around an exit/entry area of the island called the launch pad. The seals, swimming at the surface, are struck with such force that the momentum of the shark carries them both into the air in an acrobatic display that looks like a Polaris missile.

This incredible breach has been made famous on a global scale by photographer Chris Fallows (www.chrisfallows.com) and has since been showcased by numerous television production companies. These attacks, which are very short in duration, are successful about 47 percent of the time, although some individual white sharks are successful up to 80 percent of the time. This suggests that white sharks get better with more experience.

At Guadalupe Island, Mexico, where water clarity can be in excess of 100 feet, white sharks are very rarely observed attacking and killing seals and sea lions. It is thought that the exceptionally clear water forces

A white shark strikes a seal at the surface off Seal Island, South Africa.

white sharks to remain very deep in order to effectively ambush their prey. To test this hypothesis, I partnered with the Woods Hole Oceanographic Institution and Mexican scientist Mauricio Hoyos with funding from Discovery Channel to test new underwater drone technology that would follow and film white sharks around Guadalupe. Dubbed "Shark Cam," this autonomous underwater vehicle (AUV) was equipped with multiple video cameras, and able to locate, track, and videotape the behavior of tagged white sharks. What we learned was truly amazing! When Shark Cam was down about 300 feet deep, it was attacked repeatedly by white sharks with incredible speed and force. When the sharks physically attacked and bit the AUV, the force was so

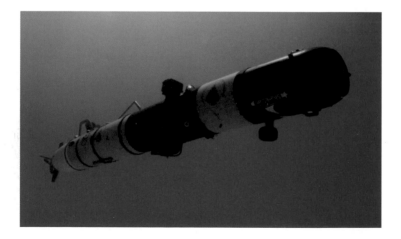

Shark Cam

White shark attacks Shark Cam

great that it pushed it as much as 8 feet vertically in the water column, leaving tooth rake marks and, in one case, compromising the hull. The vertical approaches were rapid and from depths well below the AUV. It was clear to us that the sharks were moving vertically towards the back-lit silhouette of the AUV just as they would feed in shallow water. After brief periods of following Shark Cam, some of the sharks actively swam downward, which is indicative of an effort to remain concealed at depth. At the time, these were the first subsurface observations of white shark predatory behavior at depth, and we concluded that white sharks at Guadalupe Island take advantage of great underwater visibility to search for seals in deep water adjacent to seal colonies so as to ambush and feed on them coming and going from the island.

Off Cape Cod, Massachusetts, white sharks hunt gray seals along the shallow, often murky, eastern coastline during the summer and early fall. The area spans about fifty miles north to south and it's peppered with small islands, sandbars, fast currents, and heavy surf. This is a challenging environment for any animal trying to feed, let alone a 2,000-pound white shark. Although there are hundreds of gray seals and dozens of white sharks in this area at any given time, we have observed only a handful of predation events. Unlike South Africa, California, and Mexico, this aggregation site does not have deep water close to shore, so it appears white sharks hug the bottom and ambush the seals in the shallows from the side, using low visibility to their advantage.

To better understand white shark predatory behavior in Cape Cod, we have been deploying the latest tagging technology. Called Acceleration Data Loggers, these tags contain camera systems and sensors that measure and archive the shark's tailbeat frequency, body posture, acceleration, depth, and temperature at a frequency of 20 times per second.

An ADL tag on a white shark.

A white shark attempts to feed on gray seals in the shallows off Cape Cod, MA.

These data paint a very detailed picture of the shark's behavior, not every day, but every second, so we can look for swimming patterns. The video collected by the tags allows us to observe what the shark is actually doing. So much data and video are collected by these tags that they need to be retrieved, so they are typically only deployed for a day or two before they detach and float to the surface. We're using these tags on white sharks off the coast of Cape Cod to better understand feeding behavior without having to follow the sharks day and night. We have not only directly observed white shark predatory behavior but have also seen them resting in the current and checking out floating objects.

WHITE SHARKS AS PREY

For almost every living creature, there is another living creature that will eat it. Although the white shark is at the very top of the food web, it is no exception. Remember, white sharks are not born as apex predators. They may be large at almost five feet long, but still small enough to be eaten. Given their size, young white sharks likely fall prey to other shark species, but it is not well documented. However, even large white sharks are not immune from being preyed upon. The killer whale, *Orcinus orca*, is known to feed on subadult and adult white sharks. They target the shark's liver, as several white sharks have been found dead with their livers missing. Remember from Chapter 3 that the white shark's liver can account for more than 20 percent of its body weight and it is mainly composed of oils. Therefore, the liver is an energy-rich meal for the killer whale. But white sharks seem to know when there is a killer in their midst because they leave aggregation sites when orcas are about. Off California and South Africa, it has been well-documented that white sharks leave these areas for weeks to months. I can't say I blame them.

The ocean is full of parasites, and white sharks are in no way exempt from playing host to dozens of species. There is almost no part of the white shark, inside or out, that is not vulnerable to parasitic infestation. There are small shrimp-like parasites, called *copepods*, that are perfectly adapted to securing themselves to the denticles of sharks. These parasites also live inside the mouth, swim up into the nasal cavities, and hang onto the gills. Other small parasites include tapeworms, called *cestodes*, roundworms, called *nematodes*, that infest the digestive tract, particularly the spiral intestine, and leeches. Sea lampreys also parasitize white sharks—I've seen 1- to 2-foot-long sea lampreys clinging to the pelvic region of white sharks, feeding on skin and muscle, creating large white patches. Parasites have also been found in the muscle, liver, and kidneys of white sharks. Despite the downright

Parasitic copepods on the skin of a white shark.

A sea lamprey clings to the second dorsal fin of a white shark.

ugliness of parasites, it is not in their best interest to do mortal harm to their hosts. Why kill the hand that feeds you? Therefore, white sharks are not likely to be killed by their parasites. But that's not to say that sharks wouldn't mind getting rid of them. It has been suggested that some white sharks breach to shake off parasites, but this is very difficult to prove.

Other fishes that associate with white sharks include remoras, often called shark suckers, pilot fish, small jacks, and larger schooling fishes. Remoras have a modified head that allows them to latch onto the surface of white sharks and hitch a ride.

A remora clings to the chin of this white shark.

Pilot fish and small jacks swim close to sharks and are so named because they look like they are providing guidance. These species are not parasites. Instead, they opportunistically feed on the shark's "table scraps," much like the way your dog snatches up food that drops to the floor. I've also observed small schools of larger fishes, like striped bass and bluefish, following white sharks, taking turns rubbing their bodies against the rough skin of the shark. We think these fish are using the white shark as an exfoliator to remove their own parasites on their skin.

These pilot fish are hoping for a free meal.

REPRODUCTION

The reproductive biology of the white shark remains shrouded in mystery. Remarkably, we know very little about this amazing animal when it comes to how, when, and where it makes baby white sharks. However, based on a few specimens, tagging data, and direct observations, we know some aspects of their reproduction. Scientists use a variety of methods to study reproduction in sharks, including the examination of reproductive organs and hormones, tagging, and direct observations. Most of what we know about white shark reproduction has come from using these methods.

MATURITY

Based on the examination of the white shark's reproductive organs, it is possible to determine the size at which males and females reach sexual maturity. In the male, for example, the claspers, which are inserted into the female during mating, are relatively small in specimens less than 9 feet long. However, the claspers grow rapidly as the shark grows from 9 to 12 feet in length and the sharks begin to produce spermatophores. Based on these observations, male white sharks are mature when they reach lengths of 11 to 12 feet.

The claspers of a mature male white shark.

For females, the rarity of large specimens has made it very difficult to determine size at maturity. Over the course of history, there have been only fifteen reports of pregnant female white sharks or their embryos and only a handful have been examined by scientists. Most of these sharks were captured off the coasts of Japan, Australia, or New Zealand. Based on these accounts, we know that white sharks can give birth to up to fourteen young, and size at sexual maturity for females is about 15 feet in length, much larger than the male.

MATING

Remember, sharks in general reproduce very differently from most other species of fish. Although reproductive mode can vary depending on species, ranging from egg-laying to giving birth to live young, all shark species must mate because egg fertilization is internal. This differs from most other species of fish, which release their eggs and sperm into the water column, as fertilization is external. The fact that many fishes form schools certainly facilitates spawning, but most sharks do not form schools. Therefore, they must find a mate when the time is right. The white shark is no exception.

Mature male and female white sharks likely move to very specific areas at a certain time of the year to mate. Some researchers believe that this happens when both sexes are visiting aggregation sites to feed on seals and sea lions. The best way to determine when and where white sharks mate is to collect direct observations of mating behavior. While this has been possible for many fish species, mating behavior has rarely been observed in sharks. The nurse shark is one of the few species of sharks ever observed mating. In the Florida Keys, for example, male nurse sharks move into the shallows to seek out and mate with females. In essence, the male, sometimes with the help of other males, holds the female tightly by taking her pectoral fin into his mouth, then by twisting his body around her, he is able to line up and insert a clasper into her cloaca for fertilization. Although this has never been observed by scientists in white sharks, there are a couple of anecdotal stories that might have detailed mating. In southern New Zealand, for example, a fur seal observer with the New Zealand Department of Conservation made the following observation in November 1991:

"I have unwittingly been fortunate to witness a mating (between two white sharks). I had thought at the beginning they were fighting as one animal appeared to be attempting to grasp the other with its great mouth, making great gouges in its side. However, they had eventually become motionless, one under the other, turning over from time to time, belly to belly. This obvious copulation lasted some forty minutes before the animals finally parted and glided off in opposite directions."

Several years later in the same region at the same time of year, local fishermen observed two large white sharks, belly to belly and head to head, rolling slowly and continuously on a sandy bottom in about 14 feet of water. They concluded that these two sharks must have been mating.

Without more direct observations, scientists have looked for other evidence of mating in this species, including physical signs like the presence of spermatophores in the male's claspers and mating scars on females. It is highly likely that white shark males grasp onto the fins and head area of the female in order to successfully insert a clasper, just like nurse sharks and other species. Based on these observations, some researchers feel that mating does indeed occur at seasonal aggregation sites. What are presumed to be mating scars have been observed on females off Guadalupe Island, Mexico, and we have seen them as well in Cape Cod, Massachusetts. The general conclusion, therefore, is that white sharks might mate when they are at these locations during the late summer and fall months.

These scars might be the result of mating.

However, not all researchers agree that white sharks mate at these hot spots. Using behavioral data derived from satellite tags, some believe that offshore migration is driven by the reproductive cycle. For example, the seasonal movement of adults from California and Mexico to the White Shark Café in the Eastern Pacific (see page 90) is thought to be for mating purposes. These researchers found that male white sharks exhibited a pattern of diving rapidly (almost 2 feet per second) up and down between 100 and 650 feet deep when in the Café during the spring months. This behavior, called *Rapid Oscillatory Diving*, is thought to be associated with aggregations of males that females briefly visit to select a mate. This kind of mating system, called *lekking*, is typical in a lot of different animals, including mammals, reptiles, and fish, and could be happening in white sharks. If this is the case, white shark mating occurs during the spring when they are in these offshore areas.

Collectively, the best available data, which include direct observations, tagging data, and physical signs of mating, do not give us conclusive information on how, when, and where each white shark population mates. Depending on the location, it could occur from the spring into the summer and, perhaps, as late as the fall.

GESTATION AND BIRTH

Much of what we know about white shark development in the womb comes from the examination of a few pregnant females. As you know from Chapter 2, the white shark exhibits aplacental viviparity, also called ovoviviparity, which means that there is no direct connection, or placenta, between the mother and her young. Once fertilized, the eggs hatch and develop within each uterus over the course of several months—this is the gestation period. Initially, the tiny embryos are reliant on nutrition from their yolk sacs, which are gradually depleted. They then transition to the consumption of uterine milk, an oil-rich fluid that is secreted by the uterine wall. As the young get larger, their dependence on uterine milk decreases and they start to consume unfertilized eggs continuously produced by the mother's ovary—a process referred to as *oophagy* ("egg eating"). They puncture the eggs using multiple rows of tiny teeth, which are routinely shed throughout development. During this period, the embryo's stomach becomes grossly distended, somewhat like a well-developed beer belly.

White shark embryos develop yolk sacs like the one on this embryonic mako.

As the embryo nears full-term, the distended belly diminishes, and the young shark more closely resembles its mom and dad. The gestation period of any shark is difficult to determine unless you can sample multiple pregnant females over the course of embryonic development, which has not been done for the white shark. Most researchers agree that the gestation is over a year, likely around eighteen months, and could be as high as twenty months, but this remains unknown.

White sharks give birth to, on average, five to ten pups,

A newborn white shark

but as many as seventeen have been reported from a pregnant female. Newborn white sharks are 4 to 5 feet long. It is unknown where pregnant females give birth to their young for each population. Tagging data in the Northeast Pacific suggests that large females remain offshore during gestation but return to nearshore areas to give birth to their young in the spring and summer. The capture of near-term pregnant females close to shore in other parts of the world supports this idea. This makes sense because a female carrying young likely wants to get away from those pesky males. In the Northwest Atlantic, no pregnant females have ever been examined, so where they give birth is unknown. It is generally thought that white sharks give birth to their young in the late spring to mid-summer (May to August) in the Northeast Pacific, in the spring (April to June) in the Northwest Pacific, and in the austral summer to mid-fall

(December to May) off South Australia. However, without further observations, this information is highly speculative.

NURSERIES

There is no parental care in any shark species and the newborn pups are on their own immediately after birth. Although white sharks are born at a size much larger than most fish species, they are still vulnerable to predation in their early years of life. In fact, the biggest enemy of a young white shark is a larger shark. Life can be brutal in the shark world. To maximize their survival, white sharks are either born in or move to coastal nursery areas during their first few months. The idea is to go to an area that provides ample food while reducing the risk of being eaten. Although the mother doesn't protect her young whatsoever, by pupping in or around a nursery area, she is giving her pups a much better chance of survival. Since we don't know where white sharks give birth to their young, we can't confirm that they are born in these nurseries. Nonetheless, it appears that white sharks consistently use specific areas during their first few years of life.

By far, the best way to discover white shark nursery areas is to look at where the smallest white sharks have been captured by fishermen for each population. In the Northwest Atlantic, for example, the primary white shark nursery is thought to be the New York Bight. These are the coastal waters extending from New Jersey north to Long Island. We have also encountered very young white sharks off the coasts of Rhode Island and

Massachusetts. These near-shore areas have abundant resources for the young sharks to feed on. In the Northeast Pacific, newborn white sharks live in a nursery that extends from the coastal waters of southern California south to the western coastline of Baja California, Mexico. In Australia, the coastal waters of the eastern shoreline appear to provide important nursery habitat for the white shark.

WHITE SHARK SIZE

White sharks get big! In fact, the white shark is the largest predatory shark in the world. Just how big do they get? Hollywood films want you to believe that they grow to 25 feet, even 30 feet long, and historical accounts have gone as high as 36 feet! It's not unusual to find reports of white sharks 20 to 25 feet through the nineteenth and twentieth centuries. But let's get real. Scientist have gone back and examined the jaws of some of these specimens and based on tooth size have concluded that none of them were close to 20 feet, let alone 36 feet. In reality, the largest reliably measured white shark was taken in the East China Sea and measured 19 feet, 9 inches. The next largest specimen was captured in Western Australia and measured 19 feet, 8 inches. For many years, the largest white shark was thought to be a massive specimen landed in Cuba in 1943. It was originally reported to be 21 feet in length, but subsequent analyses cast doubt on this report and its actual length is now thought to be 19 feet, 6 inches.

I have personally seen and examined white sharks in the 16 to 18-foot range off the northeastern coast of the United States and these animals look massive, so I can see how shark size can be overestimated. I personally tagged and swam with a giant white shark with my Mexican colleague Mauricio Hoyos off Guadalupe Island that he named Deep Blue. This shark was sighted again off Hawaii and has become a social media phenomenon. We thought she was in the 20-foot range, but for obvious reasons we could not get an accurate measurement. What is most striking about large white sharks is their girth. The expression that big white sharks look like small buses is not too far from the truth. As an example, I measured the girth of a 13-foot specimen to be about eight feet around and a 16-foot specimen at 9 feet around. Imagine the width of a 19-foot specimen!

The most impressive feature of a large white shark is its girth.

Of course, the best way to account for length and girth is to look at a white shark's weight. I've seen 15- and 16-foot white sharks in the weight range of 2,400 to 2,800 lbs. Given the growth in girth, weight increases very quickly with each foot in length. A 17-foot specimen taken off Long Island weighed over 4,000 pounds. The largest white shark measured at over 19 feet was thought to be about 5,500 pounds, or more than two and a half tons. You can see why they call them small buses!

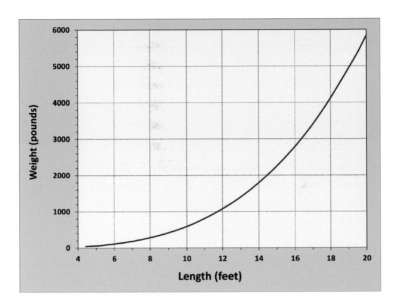

The relationship between white shark weight and length.

AGE AND GROWTH

When you think about these massive white sharks, you can't help but wonder about their age. We already know that we need to tag it and track it to figure out where it goes, how it spends its time, and how fast it swims. We know now that we can check its stomach to find out what it eats, which includes schooling fish, squid, seals, and other marine mammals, and whatever it can scavenge, like dead whales. We also now have rough estimates of size at maturity for white sharks. Now, how do we figure its age, how fast it grows, its age at maturity, and how long it lives? Well, there are only a couple of ways to estimate the age and growth rates in sharks, and all have been applied to the white shark.

The most obvious method would be to keep a white shark in an aquarium and see how fast it grows and how long it lives. However, most sharks, including the white shark, either don't survive at all in captivity or can only survive for short periods. In addition, growth in captivity tends to be a lot faster than in the wild because sharks are fed more frequently. The Monterey Bay Aquarium (MBA) in California is the only aquarium in the world to successfully maintain and display white sharks in captivity, where they were studied by researchers. The white sharks that survived in this aquarium were all very small and young, great care was taken to transport them properly, and they were kept alive in MBA's massive million-gallon exhibit. This aquarium was so large that they could also keep other large fishes, like yellowfin tuna and hammerhead sharks. Although these baby white sharks started small at only 4.6 to 5.4 feet, they were fed well and grew rapidly on a pace that would have been up to 2.6 feet per year. However, they grew so fast that they could only be held for 198 days because they outgrew the tank and needed to be released. This growth rate is thought to be roughly twice as fast as what we see in wild white sharks and a good example of how captive animals don't reflect what is happening in nature. Having learned so much about this species in captivity, MBA no longer keeps white sharks on display.

To better estimate growth in the wild, another good method is to use tagging. Imagine if we tagged and released a white shark and measured it at about 5 feet when we did so. Then the shark was recaptured five years later and it was 8 feet long. From this single tag recapture, we can estimate growth rate and a minimum age. The shark grew 3 feet in five years (7 inches per year) and was at least 5 years old. This method works well for species in which a lot of sharks are tagged and recaptured, like blue sharks. Unfortunately, not many white sharks have been tagged and recaptured in most places around the world. For example, as of 2019, only fifty-five white sharks have been conventionally tagged in the Atlantic Ocean, with only two recaptures. We can't get much growth information from only two fish, so we must rely on other methods.

By far, most of what we know about white shark age and growth comes from the most frequently used shark aging method, which involves the use of the shark's backbone. It is thought that the vertebral column of sharks, in general, grows much like a tree, laying down rings or bands that can be counted to estimate age. So, a white shark with twelve bands in its backbone is thought to be 12 years old. While this sounds like a simple

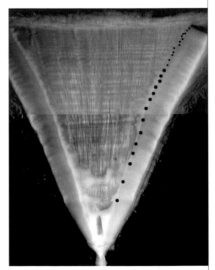

This cross section of a white shark vertebra has 27 growth bands marked with black dots.

and reliable method, sharks are not trees, and we aren't always sure that one ring is equal to one year. So, when we age sharks by counting vertebral bands, we need to validate those ages in some way. Also, you can't count the rings unless you remove the backbone, so you have to kill the shark.

Nonetheless, several researchers have used white shark vertebrae to estimate age, longevity, and growth rates in this species. The first growth curves that came out indicated that white sharks did not live much more than eighteen years and growth was relatively fast at about 1 to 2 feet per year to start and slowly decreasing over time. But these early estimates were based on a coarse examination of the most conspicuous vertebral rings and, therefore, overestimated growth rates.

More recently, researchers have been using bomb radiocarbon to validate ages derived from vertebral rings in sharks, including the white shark. Basically, this technique uses the increase of radiocarbon above natural levels that was produced as a result of atmospheric testing of thermonuclear devices during the 1950s and 1960s. This increase in atmospheric radiocarbon mixed relatively quickly into the ocean and became incorporated in the growing tissues of marine organisms through the food web. Therefore, this rapid rise in radiocarbon in the ocean can be used as a time stamp to determine the age of a shark living at the time. Working with researchers at the Woods Hole Oceanographic Institution, we examined radiocarbon levels in vertebrae taken from white sharks that were living through this period. We found out that the early growth and longevity estimates for this species were incredibly underestimated. White sharks actually live over 70 years and grow very slowly at just a few inches per year. This means that white sharks don't mature until they are into their late 20s or early 30s. We also found out white sharks don't grow much in length in their later years. When we looked at the vertebrae of the largest white sharks, we only counted up to forty-four rings, which means that just counting rings underestimates age in this species. There are also indications that females grow larger and faster than males, but more samples are needed to confirm this.

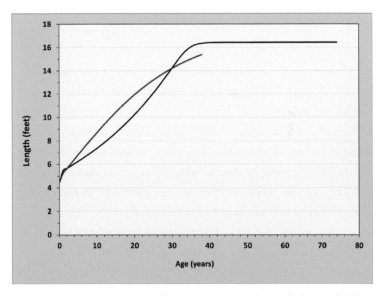

The two most recent growth curves for the white shark from the Atlantic (black line) and Indian (red line) Oceans.

The estimation of age and growth in sharks, in general, is very difficult. The white shark is a perfect example of how we need to use multiple techniques to examine this aspect of life history in these species. The most recent research now indicates that the vertebral column of sharks, including the white shark, might not be

an accurate predictor of age. In some species, the number of rings differs along the vertebral column, with more rings in the heaviest part of the sharks, the midsection. This indicates that the vertebral bands are more for structural support and not necessarily related to age! We shall see if this is the case as more research is focused on this important aspect of natural history.

Regardless, from what we do know, white sharks are among the slowest growing and longest living sharks on the planet. Like most top predators on land and in the sea, white shark populations are inherently low, so slow growth and great longevity is to be expected to enhance the survival of the species. In looking broadly at the life history of the white shark, we see a strategy emerge that is very much different from other fishes. Characteristics including slow growth, late ages of maturity, high longevity, and low numbers of well-developed young are in stark contrast to the fast growth, high numbers of larval young, and short lives of most other fishes. This strategy has worked well for the long-term survival of this species, but it also renders them particularly sensitive to any level of fishing pressure. Indeed, white shark life history is remarkably similar to another group of successful animals called mammals. It was these same characteristics that led to the depletion of whales when they were hunted to the brink of extinction in the nineteenth and twentieth centuries. Just like the whales, white shark populations have experienced declines in some parts of their range because they are so vulnerable to overexploitation.

5
CONSERVATION

O ver centuries, white sharks have been hunted by humans—sometimes for products, sometimes for fun, and sometimes just because of the misconception that they are bloodthirsty killers. Fortunately, the more we learn, the more we realize that sharks are not really what we once thought they were. They are evolutionary marvels and critically important members of the marine ecosystem. Unfortunately, these realizations have not stopped the killing of sharks. We are now dealing with the greatest level of shark exploitation by humans in history. Driven by growing populations and the fin trade, commercial landings of sharks and shark parts continues to grow. Without international management and conservation, the 450-million-year history of some shark species, including the white shark in some regions, can come to an abrupt end.

THREATS

Like all species living in the ocean during these times, white sharks are exposed to a variety of man-made threats that impact their survival. While the pollution of our coastal waters likely causes degradation of white shark habitat, we have very little data to quantify these effects. Clearly, fisheries exploitation and culling programs (see Chapter 7) have impacted global populations of the white shark and climate change has already been shown to be altering the distribution of this species.

Sharks, in general, are caught on all kinds of fishing gear for both commercial and recreational purposes. Sometimes they are targeted by a directed fishery and sometimes they are caught unintentionally as *bycatch*. Regardless, global fisheries for sharks have exploded over the last few decades. Fortunately, white sharks are protected to varying degrees in many regions, but there are strong indications that damage to some populations has already been done. In addition, white sharks are still captured

This white shark swam through a plastic strap that is now causing damage.

as bycatch and/or by shark control programs all over the world. In the United States, for example, it is illegal to possess a white shark, but many kinds of fishing gear will still hook, entangle, or trap this species.

Fishing is fun, so there are a lot of folks who go shark fishing for the enjoyment of reeling in a big fish. For much of the twentieth century, traditional big-game fishing did not include sharks; it centered on marlin, tuna, and swordfish. It was not until Hollywood portrayed the shark as a menacing man-eater in the mid-1970s that sport fishing for all sharks really took off. At that time, shark fishing tournaments, particularly off the east coast of the United States, attracted hundreds of big game fishermen each year to compete for the biggest shark. In those days, there were no regulations pertaining to sharks, and it was not unusual to see dozens of dead sharks on the dock during one of these events. The general idea was that the sport-fishing community was doing humanity a big favor by ridding the ocean of killers. These tournaments also captured white sharks, but they were far outnumbered by other species. One study showed that during a single shark tournament off Long Island, New York, from 1965 to 1983, one white shark was landed for 270 other sharks brought in. Nonetheless, of the more than 400 records of white

A white shark harpooned by a recreational fisherman.

sharks captured off the eastern seaboard of the U.S. from 1800 to 2010, 41 percent (166) were taken by recreational fishermen. In the 1980s, harpooning very large sharks also became popular and several adult white sharks were killed in this manner. For example, I had the opportunity to examine and dissect a 16-foot white shark that had been harpooned off New York in 1983. Aside from the scientific value of this shark, much of it, except for the jaws, ended up in the dumpster.

In the early 1990s, the United States government began to regulate shark fishing, but much of the data needed to do so were lacking. Given the popularity of sharks, it was remarkable that we knew so little about how they live. So, research began to really focus on shark life history, thereby shifting from the previous emphasis on shark attacks. Early regulations placed limits on both recreational and commercial fishing. In 1997, the white shark became a prohibited species on the east coast of the United States, so fishermen could no longer retain them. As we learned more and more about sharks, public attitudes started to shift away from the sentiment that "the only good shark is a dead shark." Recreational fishing ethics also changed, and more and more fishermen began to practice catch-and-release. Many joined the NMFS Cooperative Shark Tagging Program (see Chapter 3), and shark tournaments implemented high minimum sizes, bag limits, and awards for catch-and-release. Although a handful of shark tournaments remain active—after all, shark fishing is still fun—most follow strict conservation-based rules, thereby reducing the number and species of sharks brought to the dock.

COMMERCIAL FISHERIES

Sharks have been harvested for commercial products for centuries. The most obvious is meat for food, but skin has been used to make leather called *shagreen*, teeth to make weapons and tools, livers for oil and vitamins, cartilage for health pills, jaws and teeth for curios and jewelry, and fins for soup. Despite the long history of shark exploitation, it was not until the

Shark-tooth necklaces

1980s that directed fisheries for sharks exploded. For the most part, shark meat was considered unpalatable because of its high urea content (see Chapter 2). However, a decline in traditional food fisheries and the development of new techniques for preparing and preserving shark meat made it more marketable. At the same time, opening trade opportunities in China and the dramatic growth in the consumption of shark fin soup in the Far East fueled the rapid proliferation of shark fisheries throughout the world. Driven by these forces, worldwide shark landings are now thought to be in excess of 100 million sharks per year, including skates and rays. These numbers do not take into consideration those sharks killed as bycatch. The bycatch (unintended capture of a shark) may result in the shark being released alive, discarded dead, or finned. Shark finning involves the capture of a shark, the removal of its fins, and the subsequent discard of the rest of the shark. In many cases, the shark is still alive when discarded, but it will not survive for long; the shark either drowns or starves to death because it cannot swim properly.

White sharks were also captured by commercial fishermen in a variety of gear types, including longlines (fishing lines with multiple hooks set over several miles), gillnets (nets designed to entangle fish), trawls (nets dragged along the bottom), and weirs (coastal pens set up to trap fish). In most cases, white sharks were taken as bycatch in these fisheries. But, like the recreational fishery, they were taken in relatively low numbers in comparison to other species.

A white shark is inadvertently captured as bycatch in a net.

It was estimated, for example, that of the more than 105,000 sharks captured in commercial longline fisheries targeting swordfish from 1957 to 1982 in the Northwest Atlantic, only 45 white sharks were captured—a ratio of one white shark to 2,300 other sharks. Of the 404 white shark records compiled by a study we conducted in 2014, 238 (59 percent) were captured in commercial fisheries.

By the end of the 1990s, shark management measures that placed strict controls on commercial fisheries were implemented in the United States. Shark landings were reduced substantially, and the practice of finning was outlawed. Commercial fishermen also joined the Cooperative Shark Tagging Program, thereby contributing to research as well as the management process. As noted above, the white shark became a prohibited species and had to be released, dead or alive, if captured by commercial fishermen. These measures appear to be working, as some shark populations off the United States, including the white shark, are rebounding. Unfortunately, this is not the case on a global scale.

When it comes to the impacts of fishing on white sharks, it is also important to take into consideration potential indirect effects. We need to be careful not to overharvest or decimate populations of the white shark's prey species.

A white shark is released by a commercial fisherman.

We already did this once when we drove seal and sea lion populations to the brink of extinction in many parts of the world. This likely impacted the health of white shark populations and caused changes to its migratory patterns. Everything in the ocean is connected, so removing one species will certainly impact another. The white shark is no exception.

SHARK CULLS

As you will see in Chapter 7, cull programs have been used for decades to reduce the probability of shark attacks in several regions around the world. The most common methods involve the use of shark nets designed to entangle and drumlines, which are baited hooks. These methods have been commonly deployed off Australia and South Africa for many decades and have also been used sporadically in other parts of the world. While these measures have been shown to reduce the number of shark attacks in these areas, they do so by killing white sharks and many other species. Unfortunately, there are signs that these measures, which continue to this day, have impacted local white shark populations.

CLIMATE CHANGE

It is well established that global ocean temperatures are expected to rise by 2°F to 5°F by the end of this century. This might not sound like much to us, but most species of sharks live in a very specific temperature range, so this will alter their distribution. The extent to which each species reacts depends on its behavior, body size, energetic requirements, and habitat needs, as well as the rate at which its habitat is changing. For the endothermic white shark, it has been predicted that it will be only slightly impacted because of their highly migratory nature and tolerance to a wide temperature range. However, this is not the case for smaller, juvenile white sharks, which have a much narrower temperature tolerance than adults. It has already been shown that juvenile white sharks (less than 8 feet) off the coast of Central California have shifted their distribution about 140 miles north in response to warming ocean temperatures. Changes in distribution and shifts of this magnitude create potential conflicts with commercial fisheries, the conservation of protected species, and public safety.

When discussing climate change and white sharks, we must also consider impacts on their prey species. White sharks aggregate in areas of high seal and sea lion abundance. Climate change is likely to impact the distribution and reproductive success of these species. Although it has been shown that water temperature does not appear to impact white shark predatory behavior on these species, it could alter pinniped distribution and abundance such that white sharks are indirectly impacted. In addition, we know that white sharks migrate offshore to feed in deep, oceanic waters during specific times of the year, depending on location. Climate change is expected to impact the abundance of deep-water critters, like squid and fishes, which could, in turn, impact the feeding ecology and migration of this species.

As you will see in the next section, the white shark is an important part of a healthy ecosystem, so any impacts on this species could cascade to disrupt the ecosystem. Clearly, additional studies on the impacts of climate change on white sharks and their prey are warranted.

WHY SAVE WHITE SHARKS?

Many countries now realize that it is important to preserve species diversity and conserve all species at sustainable levels. Basically, that means keeping all the earth's natural resources at healthy population levels. It took centuries to get there, but we now understand that virtually all species in the ocean are interconnected. What happens to the population of one species can certainly influence what happens to the populations of the species with which it interacts. As apex predators at the top of the food web, shark populations are particularly influential in the marine ecosystem. Think about all the different species that sharks eat, for example, ranging from other top predators, like seals, to fishes and invertebrates at the bottom of the food web, like herring and squid. There are even shark species, basking and megamouth sharks,

All of the ocean's life is interconnected.

that feed on the lowest level, plankton. So, it makes sense that any impacts on shark populations will have cascading impacts on the species below them, which could be detrimental to the balance, or equilibrium, of the marine ecosystem. And we need a healthy ocean to survive!

The white shark is a perfect example of how these relationships between species are critical to ocean health. We know from Chapter 4 that the white shark is at the top of the food web, feeding on seals, sea lions, dolphins, porpoises, and other sharks in all the major oceans. As a top predator, white sharks play an important role in regulating or keeping in check the populations of these prey species. Moreover, as a top predator, the white shark's natural population size is inherently low to begin with—think of the food web as pyramid-shaped with species' densities decreasing as you move from prey at the bottom to predators on the top. You don't need to imagine what happens when you reduce white shark and/or its prey populations because it has actually happened in some regions. In the western North Atlantic, for example, humans hunted seal populations to the brink of extinction during the twentieth century. As a result, white sharks had to adapt to other prey, which changed their movement patterns and likely affected those prey species. In 1972, the United States protected seal and other marine mammal populations and, over the course of the next fifty years, populations rebounded. However, over this same period, the white shark population declined due to expanding commercial and recreational fisheries. As a result, the growing seal population grew unchecked, and people started complaining about too many seals. Fishermen, in particular, felt seals were competing with them for the same resources, namely marketable fish species. Without white sharks to maintain the balance, the equilibrium of the ecosystem was disrupted. Fortunately, white sharks were protected in 1997, and the population is rebounding and now back to hunting seals off New England. With any luck, balance will be restored.

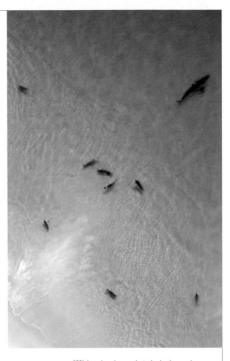

White sharks maintain balance in the ecosystem as apex predators.

The first step to knowing if conservation is needed is to fully understand the life history and population dynamics of a species. Estimates of shark populations are critical to conservation. If we don't know how many white sharks are out there, then we can't estimate the impact of fishing on populations. But we also need to know basic life history information, like age and growth, reproduction, distribution, and movements. In essence, we need to know how fast the white shark can replace itself in the face of exploitation. In Chapter 4, we learned that the white shark grows very slowly and has a low reproductive rate. Theoretically, this means that the population can be decimated very quickly when faced with fishing pressure. But how do we know how the population is trending? For that, we need to conduct a population assessment, which is not easy for any species, let alone one that travels across ocean basins.

COUNTING WHITE SHARKS

How big are white shark populations and are they growing or shrinking? We are not even sure how many species of sharks live in the ocean, so you can imagine that counting the actual number within each species is even more problematic. Unfortunately, as long as white sharks are even just interacting with fishing gear as bycatch, we need to know the answer to this question. Since it is virtually impossible to count white sharks (and any other fish species) living in the ocean, we rely quite heavily on estimating their population size using a couple of sophisticated modeling techniques.

One method is to calculate an index of relative abundance, which won't give you the population size, but might tell you how the population is trending. This is done by using catch rates from research cruises or fishing operations. For example, if a research or fishing vessel sets 1,000 hooks in the same place every year and catches 200 sharks (catch rate index = 200/1000 = 0.2) the first year, then progressively fewer and fewer sharks until the tenth year when it catches only 20 sharks (catch rate = 0.02), then this time series of data indicates that the relative abundance and perhaps the population may be declining (going from 0.2 to 0.02). For white sharks in the Northwestern Atlantic, we used data from research surveys and fishermen to show that the population declined 60 to 70 percent through the 1970s and 1980s, but has been trending upward since the 1990s. This method is used extensively for many shark species.

For white sharks, the most common method used to estimate actual population size is called *mark-recapture*, also called *capture-recapture*. This technique involves marking several individuals from a population so they can be identified later. The "marking" can be tagging or simply using features that allow you to identify them. The population is then resampled later (generally the next year), and the number of individuals within the sample that

The color patterns, scars, and fin shapes of white sharks are used to identify individuals.

are already marked is counted. This number should be proportional to the number of marked individuals in the whole population, so an estimate of the total population size can be obtained by dividing the number of marked individuals by the proportion of marked individuals in the second sample. This seems pretty simple, but the modeling is far more complicated than it sounds. For the white shark, this method has been used to estimate the number of individuals at several aggregation sites around the world. Since white sharks have unique color patterns, fin shapes, and scars, researchers can readily differentiate between individuals. So, instead of tagging them, we can catalog individual white sharks by examining photos or videos.

In recent years, another method used to model white shark populations over broader scales involves a modified version of mark-recapture called close-kin mark-recapture. Instead of using individual color patterns, this method uses genetic (DNA) profiles derived from tissue samples. The premise is the same as mark-recapture, but recaptures are determined from an individual shark's close relatives instead of that individual. Basically, DNA profiles of individuals are compared across all

A researcher videotapes a white shark for identification.

individuals to estimate abundance, movement patterns, and population trends.

There is still much work to be done and some of these population estimates are coarse at best, but they give us a sense of whether we need to do more to protect this species. Based on these methods, what do we know about white shark abundance around the world? Let's take a look.

CENTRAL CALIFORNIA, NORTHEAST PACIFIC

Based on mark-recapture, the first estimates of white shark abundance off Central California indicated that there were between 130 to 275, with an average of 219, individual subadult and adult white sharks in the region during 2006 to 2008. However, a subsequent study estimated that the broader regional population (not just subadult and adult) was in the range of 2,148 to 2,819 individuals with a point estimate of 2,418 white sharks. More recently, researchers conducted another mark-recapture study at three aggregation sites off Central California and found that the population of subadult and adults, while consistently less than 300 sharks, increased from 2011 to 2018 at these hotspots. This is good news because it shows that the protection of white sharks in California starting in 1994, coupled with bans on coastal gillnetting in 1990, are now allowing the white shark population to grow.

GUADALUPE ISLAND, MEXICO, NORTHEAST PACIFIC

Although white sharks that aggregate in Guadalupe are also part of the larger Northeast

Pacific population, there appears to be very little mixing between the populations in Guadalupe and Central California. Therefore, abundance is estimated for each region. Early counts of white sharks from 2001 to 2009 indicated that there were at least 113 white sharks at Guadalupe. A subsequent mark-recapture study based on these data produced a population estimate of about 120 subadult and adult white sharks. Another mark-recapture study conducted from 2012 to 2014 indicated that 62 to 105 white sharks interacted with ecotourism operations during that period. More recently, researchers identified 165 individual white sharks visiting ecotour vessels from 2014 to 2019. They also calculated an annual relative abundance index using the number of white shark sightings per day and found that the abundance of white sharks is increasing each year. While this is good news for the Guadalupe white shark, these researchers caution against allowing too many ecotour operators in the region.

SOUTH AFRICA, WEST INDIAN OCEAN

Based on catch rates from the KwaZulu-Natal shark nets (see Chapter 7), the white shark population appeared to remain stable from 1978 to 2003. Using mark-recapture, researchers estimated that there was between 839 to 1,843 individuals with a point estimate of 1,279 white sharks off the west coast of South Africa during 1989 to 1993. A mark-recapture study conducted 16 years later in Gansbaai produced an estimate of 908 (808 to 1,008) white sharks during 2007 to 2011. These estimates are close and clearly show that South Africa's population is larger than in other regions. However, it's clear that white shark numbers have not increased significantly despite the species being protected in South Africa since 1991. This indicates that shark nets, drum lines, and other sources of mortality might not be allowing for population growth and, perhaps, additional protective measures are needed.

AUSTRALIA, EAST INDIAN OCEAN, AND SOUTHWEST PACIFIC OCEAN

It has been shown that white sharks sampled off Australia have two genetically distinct population segments: an eastern population encompassing the Australian eastern seaboard, New Zealand, and the Southwest Pacific Ocean; and a southwestern population ranging from western Victoria to Western Australia. There is some mixing, but females return to their respective regions to reproduce. An analysis of catch rates from shark nets and drumlines indicates a more than 90 percent decline in the eastern population since the 1950s. The white shark was protected in Australian waters in 1999, but bycatch and shark control programs still capture this species. Researchers recently conducted a close-kin mark-recapture study on white sharks for both population segments. They concluded that the eastern population in 2017 was about 5,500 sharks including 750 adults. The southwestern population that year was estimated to be larger at about

1,500 adults, but they did not have enough data to estimate the total population size in that region. They also concluded that these populations have been generally stable over the last decade, indicating that the populations are not recovering despite protection. Even with no fishing mortality, white shark populations are not expected to recover quickly because of their slow growth and low reproductive rates. Therefore, conservation efforts need to be maintained.

NORTHWEST PACIFIC OCEAN

The white shark in the Northwest Pacific is poorly understood because the only available data are reports from fisheries captures and attacks on humans from Japan and Russia. As a result, even the distribution of this population

is not well known. However, an analysis of all available white shark records in the region indicates that the population was likely stable from 1951 to 2011. The amount of protection varies by country in this region, but none impose complete white shark protection.

CAPE COD, NORTHWEST ATLANTIC

With the exception of the Northwest Pacific, the white shark population in the Northwest Atlantic has historically been one of the least studied. That's because seal populations were driven to the brink of extinction and there were no white shark aggregations in this region. That has changed over the last fifteen years and white sharks are going back to aggregating off the northeastern United States and, in particular, Cape Cod, Massachusetts.

White sharks can now be predictably found in this region feeding on these seals, therefore providing reliable access to this species for study. As noted above, the first indication that white sharks were responding to protection put in place in 1997 comes from a relative abundance index from a series of sightings and capture data spanning more than 100 years. In this study, we showed that the population was overfished in the late 1970s and 1980s, but is now recovering. This information did not give us a population estimate, but it did show us that the regulations in place were working.

More recently, we worked with the Atlantic White Shark Conservancy to complete a mark-recapture study off Cape Cod. This study differed from other studies of this nature because we made every effort to incorporate the local

Most white shark population estimates come from hotspots.

movements of tagged white sharks in the work. This resulted in the first spatial mark-recapture study ever conducted on white sharks. Our results show that white shark abundance increased from 2015 to 2018 and 793 to 823 (800) white sharks visited the Cape Cod aggregation site over the survey period. In addition, as many as 263 individual white sharks were at the aggregation any given month during this period. To date, we have cataloged over 500 individual white sharks that have visited Cape Cod. The online catalog can be found at: www.shiny.atlanticwhiteshark. org/logbook/. Clearly, efforts to protect the white shark population are working in the Northwest Atlantic.

MEDITERRANEAN SEA

There is ample historical evidence that the white shark is found in the Mediterranean Sea, but it is considered rare and critically endangered in this region. Based on the analysis of occurrence records (sightings and catch data) spanning the period of 1860 to 2016, researchers concluded that the white shark population has declined 61 percent since 1975. It also appears that the distribution of this species is shrinking with some areas no longer seeing white sharks. These changes were likely brought about by overfishing and/or the declining populations of prey species, like bluefin tuna, pinnipeds, and whales. This is dismal news for this species. Clearly, the white shark needs additional protection in the Mediterranean Sea.

MANAGEMENT

Incorporating life history information, landings, and abundance estimates to assess the status of a species involves a lot of complex math, assuming we know all this information. Ultimately, in the ideal world, information based on assessments allows governments to set limits on commercial and recreational fishermen so that utilization is balanced with conservation, and white shark populations are maintained at sustainable levels. As you saw in the previous section, we know a lot about some white shark populations and very little about others. Almost all of them suffered overexploitation decades ago and need long-term rebuilding. Fortunately, some are stable or rebounding slowly, but others are not. Many countries do have protections in place for the white shark, in particular those that have aggregations sites, like the United States, Mexico,

South Africa, Australia, and New Zealand. In the United States, for example, shark management is the responsibility of the National Marine Fisheries Service and coastal states. The United States has implemented a number of measures to control shark fishing activities

We need healthy populations of white sharks.

and promote conservation. The white shark, and several other species, have been protected from harvesting, and finning is not a legal fishing activity. In addition to the United States, white sharks are protected by national legislation in Australia, Canada, Croatia, European Union, Maldives, Malta, Mexico, Namibia, New Zealand, and South Africa.

However, white sharks are not protected everywhere, and from Chapter 3 we know that this species does not pay attention to boundaries. So, white sharks might be conserved in one country, but harvested without limits in another. Therefore, we need an oceanwide approach to conserve white sharks. This is generally done with international agreements or treaties. Without the cooperation of all or most of the countries that catch white sharks, conservation efforts will fail. The International Union for Conservation of Nature (IUCN) has assessed the global status of the white shark and concluded that it should be classified as Vulnerable, which means that it is threatened with extinction. The IUCN further classified the Mediterranean population as critically endangered, which means there is an extremely high risk of extinction. These designations prompted the Convention on International

The white shark is protected by some countries, but not all.

Trade in Endangered Species of Wild Fauna and Flora (CITES) to list the white shark in Appendix II. CITES is an international agreement between 184 countries that works to ensure that the international trade of wild animals and plants does not threaten the survival of the species. An appendix II listing of the white shark means that it may become extinct unless trade is closely controlled. So, to export white sharks or white shark parts, a CITES permit is required. This allows the international community to keep white shark trade at low levels.

In addition, the white shark is listed on both Appendix I and II of the Convention on Migratory Species (CMS). The CMS is an environmental treaty of the United Nations with 133 participating countries. It is the only global agreement specializing in the conservation of migratory species like white sharks. The Appendix I listing means that the white shark is threatened with extinction and CMS countries must work to protect this species, its habitat, migration pathways, and any other factors that might endanger them. The CMS Appendix II listing means that white sharks would significantly benefit from international agreements for their conservation and management. Although these international agreements are collectively a great step forward, more work needs to be done to enforce existing white shark regulations, reduce white shark bycatch, and bring developing countries on board with conserving this species.

6
ECOTOURISM

Over the many years that I have been working with white sharks, I have found that there are many people who are horrified by this species, but just as many, if not more, who are truly fascinated by, or even infatuated with, it. One of the most common questions I get is "Where can I see a white shark?" Fifty years ago, that question was very tough to answer, but that has changed. White shark ecotourism has taken off all over the world and now almost anybody can see a white shark, assuming they have the money to do it. These adventures are exciting, safe, and expensive, but in most cases, worth every penny.

There was a time when the only people voluntarily swimming

White sharks are popular film subjects.

with sharks were those who were studying them, like me, or filming them. However, in the age of bungee jumping, sky diving, and extreme skiing, diving with sharks is just another adventurous thing to do. This has been facilitated by the discovery of shark hotspots, where it is economically feasible for a dive operator to consistently take clients to dive with or view sharks. The explosion in white shark ecotourism in recent decades coincides with the discovery of white shark aggregation sites. If you can't predictably find white sharks, you can't study them. The same applies to viewing and filming opportunities. In Chapter 3, I discussed white shark hotspots in detail. Large white sharks, generally greater than 9 feet, seasonally migrate to these locations to feed on pinnipeds (seals and sea lions). In short, white sharks do this because they can! They are large, powerful predators that prefer energy-rich prey and the migration to these hotspots is well worth the trip. This has allowed researchers to study this species intensively at predictable locations, so most of what we know about these species comes from these aggregation sites. These hotspots also provide opportunities for filmmakers and white shark enthusiasts. By far, most shows featuring white sharks have been filmed at these locations. But you don't have to be making movies if you simply want to experience the thrill of viewing or being in the water with a white shark. White shark ecotourism is available in almost all the white shark aggregation areas. In this chapter, I review where and when you can see white sharks in their natural environment and the pros and cons of white shark ecotourism.

WHERE AND WHEN?

There are currently six white shark aggregation areas in the world located off the coasts of California, Mexico, Massachusetts, South Africa, South Australia, and New Zealand. Each of these areas is remarkably different, but all have well developed white shark ecotourism opportunities. In some locations, you can couple your white shark experience with other local attractions. Depending on where you live, some are easier to get to than others. Keep in mind, these excursions are expensive, on the order of hundreds to thousands of dollars, and travel to these locations will add to the cost.

The view of a white shark from a cage.

Also referred to as the Devil's Teeth, these islands lie about 27 miles east of San Francisco. White sharks feed on juvenile northern elephant seals and California sea lions during the fall months of September to November. White shark cage diving operators leave out of San Francisco for one-day excursions, and travel to the Farallons is about three hours. You don't have to be SCUBA certified as air is supplied from the surface through a long hose called a *hookah*. Underwater visibility can be murky and expect water temperatures in the range of 54°F to 57°F. Operators are permitted to use a seal-shaped decoy to attract white sharks, but bait is not allowed. Folks who simply want to view white sharks from the vessel

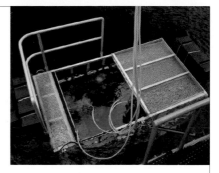

A hookah system provides surface air to divers.

and not go in the water can do so at a reduced cost. There is no guarantee that white sharks will be sighted, and the odds are considered low when compared to other locations.

GUADALUPE ISLAND, MEXICO

Guadalupe Island, Mexico, supports four species of seals and sea lions, and white sharks are abundant from July to November. There are several ecotour operators leaving from Ensenada, Mexico, or San Diego, California, for multi-day trips. Most trips are five or seven days including

Guadalupe Island

Operators are allowed to use bait, but every effort must be taken not to feed the sharks. The probability of seeing sharks is very high and most operators guarantee that your trip will be a success. I have been to Guadalupe numerous times and can testify to the fact that this is one of the best white shark cage diving locations in the world. Unfortunately, at the time of this writing, white shark ecotourism has been closed at Guadalupe because of a series of mishaps and injuries to white sharks. It remains to be seen if it will ever be reopened.

roughly two days of travel from Ensenada, Mexico to Guadalupe. All these excursions offer surface cages with hookah, so SCUBA certification is not required. Some operators also offer cages for those certified divers who want to go deeper. Water temperatures range from 66°F to 70°F and visibility is typically 100 feet or greater. July and August trips are dominated by male sharks, females arrive in September, and low numbers of very large females arrive for October and November.

WESTERN CAPE, SOUTH AFRICA

The white shark is broadly distributed in this province, but most ecotour operators work out of Cape Town, Gansbaai, or Mossel Bay. Although white sharks can be found year-round

Guadalupe Island has crystal-clear water for viewing white sharks.

in this region, peak season is during the austral late fall and winter when white sharks feed at island rookeries of Cape fur seals from May to October. White shark trips vary by location and ecotour operator. Most offer half-day excursions, but some have multi-day options. South Africa has the greatest number of cage operators in the world, so pricing is more competitive than in other areas. Dive certification is not needed because most operators, with only a couple of exceptions, use surface cages without an air supply—divers hold their breath to view the sharks. Water temperatures can range between 54°F to 68°F and visibility is generally poor, typically less than 10 feet depending on conditions. Some companies offer ride-along opportunities at a reduced rate for folks who do not want to get wet.

South Africa is famous for its breaching white sharks.

Operators out of Cape Town visit Seal Island in False Bay, which is famous for its breaching white sharks (see Chapter 4) during the peak months of June to August. This activity happens in the morning, so these vessels leave the dock early in hopes of seeing a white shark breach. The highest number of operators work out of Gansbaai, dubbed the "shark cage diving capital of the world," around Dyer Island, another Cape fur seal colony. The sharks are often seen cruising the nearby channel, called Shark Alley, on the southern side of the island. Although further east, Mossel Bay hosts a couple of operators who generally work near another Seal Island off Diaz Beach, which has a large breeding colony of Cape fur seals. Being in the bay, this location is well-sheltered and provides an excellent option for those travelling the coast.

Over the last several years, the number of white sharks has been decreasing off the coast of South Africa. This has largely been attributed to predation by killer whales (see Chapter 4). It has been shown that white sharks exhibit a flight response when orcas are about. There is also concern that local fisheries are depleting populations of small sharks, which are a major part of the white shark's diet. Although numbers are down and, in some areas disappearing, white shark cage diving is still very viable in South Africa. It is best to check with eco-tour operators regarding local abundance before committing to the trip.

In South Australia, white sharks aggregate at the Neptune Islands to feed on long-nosed fur seals. While males visit these islands year-round, large females are there in the austral fall and winter, likely to feed on lactating fur seal mothers and their pups. In Australia, shark cage dive operations work out of Port Lincoln, which is about forty miles from the Neptune Islands. White shark diving tours are offered year-round by two operators. One specializes in day-long and twilight trips to and from the islands with a surface cage and hookah breathing. The other runs multi-day excursions (three to ten days) with a surface cage for

North Neptune Island

non-divers and a deep cage for certified divers. Water temperature ranges from 57°F to 68°F depending on time of year and water visibility is excellent at about 50 to 100 feet. Operators can use chum, locally known as *berley*, to attract sharks to the vessel. There are no guarantees that you will see white sharks, but one operator boasts a success rate of 80 percent.

STEWART ISLAND, NEW ZEALAND

Although there are two predominant aggregation sites in New Zealand at the Chatham Islands, 400 miles off the east coast, and Stewart Island off the southern tip, white shark cage diving is only available at the latter. Both locations support large breeding colonies of long-nosed fur seals. White sharks aggregate at these sites from the late summer to early winter,

peaking in the fall. There is currently only one cage dive operator in New Zealand in the town of Bluff in the Southland region. This location provides easy access to Foveaux Strait and the Titi Islands, which separates the mainland from Stewart Island. This operator provides daily ecotours from December to June. Bait is used to attract white sharks and guests view the sharks from a surface cage with air supplied by hookah. Non-certified divers are given a brief training session before entering the cage. Water temperature ranges from 50°F to 68°F and visibility is good at 30 to 60 feet.

CAPE COD, MASSACHUSETTS

In the Northwest Atlantic, Cape Cod, Massachusetts, has emerged as an aggregation site for white sharks due to the restoration of the gray seal population. White sharks aggregate at Cape Cod from June to November. The aggregation attracts subadult and adult white sharks; peak months are August, September, and October. This is the location where I conduct the bulk of my white shark research with the Atlantic White Shark Conservancy. Because this aggregation is a popular vacation destination attracting tourists from all over the world to enjoy the beautiful beaches, the state of Massachusetts has made it illegal to chum or attract white sharks in any way within state waters (out to three nautical miles from the shoreline). As a result, there are no cage diving opportunities in Cape Cod. However, there are shark viewing excursions offered by several local charter vessels. These tours are two to four hours long and they originate in Cape Cod ports within minutes of the white sharks. Sharks are located by a professional spotter pilot, who directs the viewing vessel to sharks in shallow waters. Water visibility can be poor, but the sharks are easy to see and highly abundant—some operators offer a guarantee of success. For those who like to fly, some airports on Cape Cod offer aerial sightseeing tours to view white sharks from the air.

Viewing a white shark off Cape Cod.

THE PROS

As you can see, there are plenty of white shark ecotourism opportunities for anybody who has the interest, the time, and the money to go to one of these hotspots. There are a lot of economic, social, and conservation benefits to white shark ecotourism. First and foremost is the education it provides and the respect it inspires in people. I've been diving with sharks for over forty years to photograph and to study them, and I love it. Each and every

Diver observes a white shark

experience, from my first dive with blue sharks in 1983 to diving with white sharks in recent years, has been truly amazing, invigorating, and educational. Although I've been working with sharks over all these years, it never gets boring for me. So, if it is that satisfying for me, I can only imagine how amazing it is for the person who has never seen a white shark in the wild. White shark ecotour operators provide that opportunity to thousands of people all over the world every year. Some of these people walk onto the boat thinking they are going to see a "man-eater" or "demon-fish," and what they see is a marvel of evolution, a majestic animal. Studies have shown that a white shark cage diving experience results in a positive shift in an individual's understanding, awareness, attitude, and concern for sharks. This leads to more respect for the species and, ultimately, adds another voice for white shark conservation.

Cage diving inspires white shark conservation.

There are other notable benefits to white shark ecotourism beyond its educational value. There was a time when the value of a shark could only be measured at the fish market where it was sold for consumption. Unfortunately, that is still the case in many areas. However, with the explosion in shark ecotourism, many are realizing that living sharks are far more valuable than dead ones. Consider that a single white shark landed and sold will produce a finite sum of money. Now consider the value of that same shark left in the water where it might be seen by dozens of ecotours over multiple years. The benefits to the local economy are obvious, in addition to the benefits to shark populations through enhanced conservation.

White shark research has also benefitted from white shark ecotourism. In most areas, ecotour operators work closely with local researchers to collect data and identification photographs, as well as provide a platform for field sampling. Indeed, much of what we know about white sharks has come from working in these aggregation sites and, in many cases, alongside ecotourism companies. Many of these operators also donate a portion of their fees to white shark research initiatives.

With just a couple of exceptions, like Cape Cod and the Farallons, most white shark ecotour operators are allowed, to varying degrees, to use bait and/or chum (ground-up fish) to attract white sharks to their vessels. This is typically referred to as provisioning, which is providing some incentive for the sharks to be there. There is concern that provisioning leads to changes in natural behavior, like conditioning sharks to associate humans with food or altering behavioral patterns. There is also concern that white shark diving operations draw sharks into coastal waters and stimulate them to feed, thereby increasing the risk of shark attacks in nearby areas.

Using field observations and various tagging technologies, researchers are examining the potential impacts of shark cage diving on white shark behavior. This research indicates that not all white sharks react the same way to the presence of shark cages—some ignore them, and some don't. While provisioning

Some think that cage diving can draw sharks closer to shore.

has been shown to influence the local fine-scale behavior of some animals, this does not happen in all areas. For example, researchers in South Africa and Guadalupe concluded that there was no significant impact on white shark behavior during periods of shark diving. They found that white sharks that received more rewards (bait) did not stay longer, nor were they able to meet their energetic needs with offered bait. However, in Australia, the number of white sharks and the amount of time they spend in the area both increased with increasing numbers of ecotour operators. These sharks also spend more energy when in the presence of shark cages. Nonetheless, there was no evidence that these cage diving operations altered the white shark's diet and nutritional condition. Overall, most studies indicate that the impact of cage-diving is likely small if interactions with individual sharks are infrequent. In addition, it appears unlikely that white sharks exposed to cage diving activities are any more likely to present a risk to divers, swimmers, or surfers in areas not in close proximity to cage diving sites.

BOTTOM LINE

When done in a responsible manner, shark cage diving can provide a net conservation benefit for white sharks by raising awareness, instilling a conservation ethic, and generating shark data. To minimize the potential impacts of cage diving on white sharks, humans, and the local ecosystem, cage diving activities are heavily regulated at all the major hotspots. To start, white sharks are protected in all areas where cage diving currently occurs, and the use of chum and bait can be minimized because large numbers of white sharks are already aggregated to feed on pinnipeds. Other regulations, which vary with location, limit the number of vessels, restrict activity to certain locations, require the collection of data on shark activity, establish protocols that minimize impacts on sharks and the local environment, and enforce methods for the handling of baits to minimize shark feeding and contact with cages. Although there are clear examples where white sharks have been injured during cage operations, that was the result of negligence on the part of the ecotour operator. Clearly, enforcement is an important aspect to sustainable and safe white shark cage diving.

Eye to eye with a white shark on an ecotour.

7
SHARK ATTACKS

Nothing is more bone chilling than the thought of being attacked by a wild animal. Most people know they will never encounter a lion or grizzly bear, but somewhere in the back of some folks' minds is the feeling that being attacked by a shark is within the realm of possibility. This cannot be further from the truth. In fact, the probability of a shark attack is so low that I hesitate to discuss the subject. When I lecture to general audiences, I do not spend a lot of time discussing shark attacks, simply because I know that virtually everybody in the room is more susceptible to harm from their own dogs than from sharks.

The probability of a shark bite is extremely low.

Yet when I poll those same audiences, I am always surprised at the number of people who fear sharks to the extent that they will not go swimming at any beach at any time.

Media companies that sell newspapers, make movies, deliver the news, and post to social media know this. Sharks make them money and shark attacks make them even more money. You might be washing dishes after dinner with the news on in the background barely paying attention, until you hear the words "a young man was attacked by a shark today." You then turn to the TV and pay close attention. While I think the fear of being bitten by a shark can be blown out of proportion, I do think the fear of being killed by a wild animal is inherent in all of us. After all, we didn't always live in protective buildings in modern cities and towns. When you consider that the modern human species emerged about 200,000 to 300,000 thousand years ago, you realize that most of that time we were exposed to the real threat of being harmed by wild animals. That fear is healthy to some extent because it enhanced our survival. But when Hollywood taps into that fear and portrays all sharks as mindless killers and the white shark specifically as a "man-eater," reality becomes fantasy. In this chapter, we get back to reality.

THE FACTS

In the late 1950s, a group of shark researchers started the International Shark Attack File (ISAF), an authoritative compilation of all known shark attacks. With reports dating back to the late 1500s, the ISAF is housed at the Florida Museum of Natural History. To date, the file has compiled over 6,800 shark attack investigations, each based on eyewitness accounts, newspaper articles, and/or photos. This is the most comprehensive shark attack database in the world and anybody can access the statistics on their website: www.floridamuseum.ufl.edu/shark-attacks/.

When we look at the last 100 years, the long-term trend indicates that the number of shark attacks has increased from less than forty during the first decade of the twentieth century to almost 800 during the most recent decade (2010 to 2019). ISAF researchers, though, are quick to point out that this rise is likely associated with an upsurge in the number of people going into the ocean and not an increase in the rate of shark attacks. People are drawn to the shoreline—just look at any travel website and you will see countless images of white sandy beaches promising relaxation. Throw in the dramatic rise in the human population and all the exciting water sports that have developed over the last century, like scuba diving, surfing, and spearfishing, and you have millions more people in the water than a hundred years ago. For example, the ISAF estimates that more than 118 million people went to United States beaches in the year 2000, resulting in 3,300 surf rescues, yet there were only 41 shark bites. It only makes sense that as more and more people go into the ocean, which is where sharks live and feed, the probability of a shark encounter is likely to increase. This growth in shark attack reports is also related to the fact that the ISAF has gotten much better at tracking shark attacks. In the early part of the last century, there was no ISAF, so they had to rely on published reports during a time when communication was nothing like it is today.

From 2012 to 2021, 761 shark bites were recorded by the ISAF worldwide. Most of these (34 percent) were reported from the state of Florida, but others occurred frequently in Australia (19 percent), Hawaii (10 percent), South Carolina (6 percent), North Carolina (4 percent), South Africa (4 percent), and California (4 percent). The number of shark attacks fluctuates each year due to annual variability in environmental, economic, and social conditions that influence the distribution of sharks and the number of people going in the water. The five-year global average from 2016 to 2021 is 72 unprovoked incidents per year. The 2021 worldwide total was 73 confirmed unprovoked cases and 9 fatalities, which is higher than the global average of 5 unprovoked fatalities per year. Despite this slight increase in 2021, fatalities are trending down over the long-term.

According to the ISAF, most victims are surfers, followed by swimmers and then divers. It is important to note that by far most of these attacks are minor, resulting in flesh wounds and

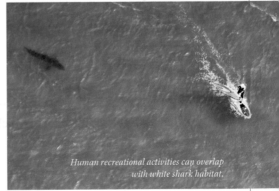

Human recreational activities can overlap with white shark habitat.

abrasions that are quickly and easily treated. This is further supported by the fact that only about 7 percent of all shark attacks reported to the ISAF since 2010 have been fatal. When you consider that over 45,000 people die in car crashes and 60,000 people die from the common flu every year in the United States alone, you can see that shark attack is not the reason to run out and buy life insurance. Putting this into the context of beach-related fatalities, the ISAF concluded that the odds of suffering a drowning or other beach-related fatality is about 1 in 2 million, whereas the risk of being attacked by a shark is about 1 in 11.5 million and the odds of a shark-inflicted fatality is about 1 in 264.1 million.

THE CULPRITS

It is extremely difficult to positively identify the species involved in a shark attack, but the ISAF has been able to do so in more than 800 cases. If we only examine those that were positively unprovoked (about 50 percent), during which the victim did not instigate the attack in any way, then the most frequent species implicated in the attacks are: great white shark (37 percent), tiger shark (15 percent), bull shark (12 percent), several requiem (*Carcharhinus*) species (blacktip, spinner, oceanic whitetip, dusky, sandbar, blacknose, Galapagos, silky,

White sharks have been implicated in more unprovoked shark bites than any other shark species.

blacktip reef, grey reef, Caribbean reef, unidentified, 18 percent), and the balance (12 percent) associated with nineteen other species. ISAF personnel acknowledge that the list may favor those species that are easy to identify, but there is strong evidence, nonetheless, that the white, tiger, and bull sharks are the top three species most likely to be implicated in attacks on humans.

The tiger shark is second on the list of shark culprits.

The bull shark is third when it comes to unprovoked shark bites.

THE "MAN-EATER"

With 351 documented interactions (51 fatal), the white shark is the number one shark species implicated in attacks on people. As a result, it is often called the "man-eater," although if humans were really part of its diet, there would be countless more attacks. Besides, why would a shark that evolved and lives in the ocean come to rely on animals that evolved and live on land for sustenance? No species could survive if its food source lived in a completely different environment. So, why do white sharks attack people? Unfortunately, the study of shark attacks is a largely inexact science because it is virtually impossible to determine the motives for such attacks. In almost all cases, the perpetrator of the attack is not apprehended and, in those cases when it is, it's simply not talking. Scientists can only really examine the various aspects and details of the attack and compare them to what is known about the behavior, life history, and ecology of the shark species that is implicated. What do we know about the life history of white sharks that would help us to determine why this species is occasionally, albeit rarely, motivated to bite humans? In looking at the diet and feeding ecology of the white shark, we know it feeds on large prey, including marine mammals, when they attain sizes greater than nine feet. We know that the white shark ambushes its prey with speed, stealth, and power typically producing a large wound, which causes the

victim to bleed to death. This information has led researchers to conclude that attacks by the white shark are likely associated with mistaken identification for different prey, namely a seal, sea lion, turtle, or dolphin.

A swimmer at the surface can be mistaken as a porpoise by the white shark.

Research has shown that the mistaken identity theory is likely the cause of most shark attacks when you take into consideration the visual system of the white shark (see Chapter 2). We know white sharks are color-blind, which means that they rely heavily on motion and contrast as the primary visual cues to detect and target prey.

Like most sharks, white sharks also have poor spatial resolving power, so they have trouble distinguishing fine details in an image. These visual constraints make it difficult for white sharks to differentiate between the shape of surfboards, human swimmers, and natural prey, namely seals.

The silhouette of this seal is very similar to that of a surfboard.

Attacks by white sharks have occurred all over the world but are clearly more common in areas where they aggregate seasonally like off California, the northeastern United States, South Africa, Australia, and New Zealand. These are areas where white sharks hunt seals and sea lions close to shore, which coincidentally overlap with popular human activities like surfing, swimming, and diving. It has also been shown that juvenile white sharks in the size range of 8 to 11 feet are responsible for most of the bites on humans. It is possible that white sharks in this size range are just beginning to incorporate seals in their diet and, therefore, are

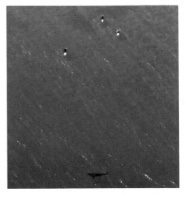

A white shark swims just outside the surf zone.

more likely to make mistakes. The fact that most victims are not actually consumed suggests that the shark may realize that it has made a mistake, based on taste or texture, and abandons the person. Unfortunately, a bite from a white shark is significant and results in victim mortality about 20 percent of the time.

AVOIDING SHARK ATTACKS

In my opinion, the best way to avoid shark attack is through education. People need to know that the risk of being attacked by a white shark is negligible, except in places where they aggregate to feed on seals. Folks who want to swim in the ocean should swim in the ocean, but do a little homework beforehand. First, know your own swimming strengths. By far, more people die from drowning than from shark bites. After all, we are land animals and the very best of swimmers can still drown under certain conditions. Therefore, make yourself familiar with local conditions. Is the area prone to dramatic tides, heavy currents, and deep drop-offs? These are all conditions that can impact a swimmer. Next, if you are concerned about dangerous wildlife, do some research, or ask local officials about what species live in the area.

I have personally been stung by a lot of jellyfish, including Portuguese man o' war, over the years, but never attacked by a shark. I wish I'd known about those jellyfish! If there are sharks in the area, what species? Have they been implicated in unprovoked attacks? Ask the local experts. In Cape Cod, we make every effort to share what we've learned about white sharks with local safety officials and the public. Attacks by white sharks tend to occur where they aggregate to feed on seals. If white sharks are documented in the area, see if there are also seals and or sea lions. Look for flocks of birds actively circling

Warning signs and flags are used in some areas to notify beachgoers about sharks.

because they are good indicators of feeding sharks—they pick up the scraps. By taking the time to assess the environmental conditions and the local wildlife, you can make an informed decision on how, when, and where you want to swim in the ocean. So, if you are in an area where white shark attacks have been documented, and there are plenty of seals or sea lions, consider staying in shallow water or avoiding the area altogether.

White sharks are more likely to attack individuals, so swimming in a group is also a good idea—this is a good safety measure regardless. Stay close to shore, do not swim with an open wound, and avoid wearing shiny jewelry. Don't swim during periods of low-light levels, like dawn and dusk, because some white sharks are more active during these periods. Murky waters and deep troughs between sandbars are good feeding areas for white sharks because they can ambush prey, so don't put yourself in the position of looking like prey. Finally, don't harass, bother, touch, or grab a shark in any way—this can provoke an attack. Although this seems like common sense, many a fool has done it.

In addition to these recommendations, researchers are now using tagging data to get a better sense of when, where, and under what conditions white sharks are most likely to be feeding. Remember, we learned in Chapter 3 that the movement patterns of white sharks can be studied using a variety of tag technologies. Specifically, acoustic transmitters coupled with receiver arrays (for detecting the sharks) tell us a lot about fine-scale movements, residency in specific areas, and seasonality. Using this technology, we can get a sense of the environmental conditions that drive the presence of white sharks near popular beaches. For example, we have been tagging white sharks off the coast of Cape Cod for several years using this technology. We've learned that white sharks arrive to feed on

gray seals as early as May and stay a late as December, but peak months of abundance are August, September, and October. We also know now that although white sharks travel throughout the coastal waters of Massachusetts, the highest densities are off the eastward

Data from tagged sharks can inform beach managers.

facing beaches of Cape Cod. Taking it a step further, we examined the effects of water temperature on the presence of white sharks and found that they were more likely to be present when water temperatures were greater than 52°F. Collectively, these results indicate that the risk of shark attack is greatest off these beaches specifically from August through October, but also depending on water temperature. This is useful information for beach managers as well as the public. Other researchers have done similar analyses in other parts of the world where white shark attacks are relatively more common.

EARLY WARNING SYSTEMS

Some areas are now deploying methods to detect the presence of white sharks and warn beachgoers. The Shark Spotters program in South Africa, for example, places spotters at high points of land near popular swimming beaches. When they spot a white shark, they sound a siren and people are notified to leave the water. Since 2004, Shark Spotters have reported more than 2,000 white sharks at South African beaches. Other early detection systems include the use of drones to spot sharks as well as underwater sonar systems to detect their presence. Both of these methods can be effective, but only under the right conditions.

Shark spotter looking over a swimming area

Glare, low light levels, underwater noise, poor visibility, and other environmental conditions can change dramatically from place to place, day to day, or even over the course of the day. We've been using drones to observe white shark behavior, for example, and found that water turbidity and time of day impact our ability to see white sharks. Therefore, these methods need to be vigorously tested to evaluate their effectiveness in each area deployed.

Off Cape Cod beaches, we are currently deploying acoustic buoys that transmit the detection of tagged white sharks to lifeguards and beach managers in real time. So, any tagged white shark that passes within 300 to 500 yards of the receiver will trigger a message to public safety officials, who can then pull swimmers from the water. Those detections are also linked to the Sharktivity app developed by the Atlantic White Shark Conservancy so the public can track the presence of these sharks. This system works well, but we do not encourage people to use it strictly for public safety. Instead, we see it as a way to educate safety officials and the

Real-time acoustic receiver used to inform lifeguards about the presence of tagged white sharks.

public about the presence of these sharks. It is important to emphasize that these systems do not detect sharks that are not tagged, and we certainly have not tagged them all. So, just because a white shark is not being reported at a beach does not mean an untagged shark is not there. The bottom line with any of these measures is that none of them are foolproof early detection methods and many, therefore, can give the public a false sense of security. Knowing the limitations of each system in each location can go a long way in educating the public and building confidence.

SHARK CONTROL

The historical approach to avoiding shark attacks has been to implement shark control programs. The two most common methods involve the use of barrier nets, typically referred to as shark nets, and drumlines (baited hooks). Shark nets and/or drumlines have been deployed off Australia since 1937, South Africa since 1952, and intermittently in other parts of the world in response to perceived increases in shark attacks. Despite how they are typically viewed, these very large nets do not create an exclusion barrier between swimmers and sharks, which means that the sharks can swim around them. Both shark nets and drumlines are designed to intercept and kill sharks before they get to the beaches so as to reduce the chances of shark attack. Studies have shown that these concerted measures have reduced the incidence of shark attack in these areas but have not completely irradicated it. However, critics are quick to point out that shark nets and drumlines have bycatch, meaning they indiscriminately kill all kinds of marine wildlife including protected species. Indeed, during the period of 2012-2021, shark nets off 51 beaches in New South Wales, Australia, killed 1,344 sharks (18 species including 116 white sharks), 339 rays (11 species), 83 turtles (5 species), 54 dolphins (3 species), and one humpback whale. Many of these animals were captured on the beachside of the net, which means that the sharks already had free access to swimmers. Despite the use of these nets, there has been 36 unprovoked shark interactions with swimmers since 1937. So, there is a balance to be struck between shark control programs and the ramifications of bycatch to shark populations and other marine species. Although efforts are being made to reduce this bycatch, nonlethal deterrents are needed.

In some cases, directed shark culls have been used to seek out and kill so-called "problem sharks" in response to shark attacks. However, there is no evidence that these shark hunts are effective at achieving their goals. For example, after seven fatal attacks caused by white sharks from 2010 to 2013 in Western Australia, the Fisheries Minister hired commercial fishermen to kill sharks more than 10 feet long. The cull was initiated in 2014 using drumlines and resulted in the capture of 172 sharks, most of which were tiger sharks—no white sharks were captured. White sharks, and other species, can move great distances over short periods of time, so the likelihood of catching the shark that attacked a human is extremely low. For this reason, shark culls do little more than kill "innocent" sharks and provide the public with a false sense of security.

With new technologies emerging almost every day, there are alternatives to lethal measures. Several companies are now marketing products that can be used to deter sharks without killing them.

BARRIERS

More sophisticated barrier systems that don't entangle animals are being deployed in some areas. For example, Shark Spotters in South Africa use an exclusion net with very fine mesh to completely cordon off an area for swimmers. The net is placed and hauled every day and the mesh is too fine to ensnare sharks and other marine animals. Another new barrier system involves the use of multiple rows of hinged PVC piping anchored to the seafloor. Rare-earth magnets, which create a powerful electrical field, are mounted on the barrier as well. The principle behind this design is that the barrier and the magnets create visual and electromagnetic deterrents, respectively. These barriers don't capture or kill any animals and have proven to deter sharks in some areas, but they are not 100 percent effective—some sharks penetrate the barriers. Barrier systems can work well in relatively shallow, sheltered areas not exposed to heavy surf, big tides, and frequent storm events. However, anchoring and maintaining any barrier in a highly dynamic environment is costly and not likely to remain in place.

A number of personal deterrents are now available for swimmers, surfers, and divers to wear or attach to their equipment. Almost all of them generate an electrical field, which is thought to overwhelm the shark's electrosensory system (see Chapter 2) and repel it. There are several configurations commercially available, and the size of the electrical field depends on the design. Small units generate relatively small fields that do not cover the entire body, while those mounted on surfboards increase the size of the field for more protection. A number of tests have been conducted on the various designs and, with very few exceptions, none of them significantly deterred white sharks from taking baits. The best performance reduced the number of interactions by about 60 percent. Remember, electrosensory perception in sharks is a near-field sense, which means that the sharks must be very close, certainly within a few feet, to be repelled by the electrical field. White sharks attack with speed and stealth, so it is unlikely that a shark fully committed to an attack would be stopped by one of these devices.

It is thought that personal devices that emit an electrical field will deter white sharks.

It would be great if there were a device, barrier, or other technological development that is 100 percent effective at deterring white shark attacks, but as I have often said, there is no silver bullet—at least not yet. While some of these approaches give swimmers and other water users peace of mind, many also present a false sense of security.

There is only one way to avoid shark attack 100 percent of the time and that is to not go into the water. Complete avoidance is also the best way to avoid car accidents, dog bites, being hit by lightning, and so on. But the truth is we are still willing to get in cars, buy dogs, and go outside during a storm. We are willing to accept that risk, which is orders of magnitude higher than the risk of being bitten by a shark. Sharks live in the ocean, it is their world, and we are guests. I believe it is our responsibility to respect their world and modify our behavior so we can coexist with white sharks.

8
THE FUTURE

O ver history, our attitudes toward sharks have ranged from worship to irrational fear. Some early civilizations revered and respected sharks as mythological gods. The shark god, Dakuwaqa, lived long ago in the seas of Fiji and was regarded as the most ferocious of all gods and the protector of fishermen, the reefs, and the ocean. In Hawaiian mythology, there were several shark gods, like

Kamohoali'i, Kua, and Ukupanipo who were considered guardians of fishermen and the oceans. In ancient Greece, Lamia was the shark-like daughter of Poseidon, who was believed to eat children. This mythological figure would form the basis for the white shark's family name Lamnidae.

As civilization took to the sea, this reverence for sharks was replaced by fear. Sailors told tall tales of sea monsters and sharks were among them. While all sharks have been described over the course of history as ancient, mindless, man-eating, killing machines, it seems the white shark has been viewed with a particular emphasis on evil. From the earliest times, its reputation as a fearsome beast has been embellished. This was even the case in the scientific community dating back centuries. In the late 1700s, British ichthyologist (a biologist who studies fishes) Thomas Pennant wrote, "They are the dread of all sailors in all hot climates, where they attend the ships in expectation of what might drop overboard; a man who has this misfortune inevitably perishes." In 1806, the famous biologist Linnaeus described it as "the most dreadful and voracious of all animals; preys upon everything that comes its way, even its own tribe, and has been known to swallow a man whole." This portrayal did not change much over the years. In 1949, marine biologist J. L. B. Smith wrote that "this swift, voracious, and ferocious shark is a terror to all who venture on or in the water." These were scientists! You can just imagine what the rest of the world was thinking.

Through the twentieth century, the reputation of sharks did not change dramatically. Shark attacks close to home and those associated with naval disasters during World War II did little to shift public perception. Much of the research associated with sharks focused on methods to prevent shark attack. There was no internet and social media back then, so virtually everything we learned came from newspapers, television, movies, and magazines. And these media outlets knew that the fear of sharks drew an audience and, as a result, made them money. Hollywood movies were no exception and it's no secret that some of these films were blockbusters. Sharks remained enemies of the people through most of the twentieth century.

To the fishing community, sharks were in the way of catching other more enjoyable or more profitable species. For recreational fishermen, tunas, marlin, and swordfish were the preferred

gamefish, not sharks, which were viewed as the scourge of the ocean. Early fishing tournaments that targeted sharks did so to remove them, not only to make room for their preferred quarry, but also to do the public a favor. There were no markets for sharks, so commercial fishermen had no interest in landing them. They also targeted more profitable species, like tunas and swordfish, so any shark that took one of their hooks was deemed a thief and promptly killed.

While public attitudes for sharks changed little, fishing for sharks took a dramatic turn and really accelerated during the 1970s and 1980s. Recreational fishermen began shifting to sharks as their preferred species began to dwindle. Shark fishing tournaments became more popular as fishermen vied for prizes and competed to land the largest shark. The tunas and swordfish so enjoyed by the recreational fishery also produced a living for the commercial fishing industry. As those species declined, these fishermen turned to sharks. The explosion of the Asian fin market at this time caused a global demand for shark fins. People were still generally frightened of sharks, but now sharks were worth something. Fisheries began landing sharks at very high levels all over the world, including the United States.

We know from Chapter 4 that the life history characteristics of sharks, including slow growth, late ages at maturity, and low numbers of young, make their populations very vulnerable to high levels of fishing mortality. By the late 1980s and early 1990s, it was starting to become obvious that shark populations were in trouble.

The best available data, and there were not much, showed rapid declines. We didn't know it at the time, but white sharks were also being landed at unsustainable numbers—see Chapter 5. Most governments were caught flatfooted with no regulations in place to curb the slaughter. It didn't help that we knew very little about the biology of most of the species being landed, including the white shark.

By the end of the twentieth century, we knew that sharks were in trouble and many countries began to take steps to curb shark fishing mortality and restore populations. We needed more science as well because we had very little life history information for these species, so shark research also exploded. During these years, internet communication and social media blew up, for good and for bad. Among the good things was that scientists and conservation organizations began communicating with the public much better. A more educated public is one more likely to protect sharks. Yes, many people were still afraid of sharks, but others were beginning to embrace these animals as incredible, awesome creatures. Shark cage diving exploded, and more and more people began really experiencing many species of sharks, including white sharks, in their natural environment. As we learned in Chapter 5, those countries for which white sharks were valuable commodities began to protect this species.

I've always been interested in white sharks, but my research on live white sharks didn't really start until 2004. It was an event that happened that year as well as fifty years earlier that gives me a sense that public attitudes toward white sharks are changing for the good. In August of 1954, a very large white shark swam into a harbor on the small island of Cuttyhunk off the coast of Massachusetts. Within hours, local fishermen harpooned and killed the shark in the interest of public safety. Fast-forward fifty years. In September 2004, another large white shark swam into a small coastal pond on a small island just a few miles north of Cuttyhunk. The 14-foot female, eventually named Gretel, became trapped and needed to be rescued. I was the lead scientist heading up the rescue and I was convinced she would die if we didn't help. Emails and phone messages began pouring in as we struggled to find a solution. What I found most inspiring was that, with few exceptions, most people pleaded with me to save the shark. Of course, this was in sharp contrast to what had happened in Cuttyhunk fifty years earlier. It took us almost two weeks, but we were eventually able to guide her to the safety of deep water. That event taught me that conservation was beginning to pay

White shark killed in 1954.

off not only with bringing back white sharks, but with changing the way the public perceives this species.

Unfortunately, given the slow reproductive rate of white sharks and other sharks species, it takes a very long time to restore shark populations once they are diminished—not years, but decades. And we've also learned that restoring the white shark population comes with potential repercussions, as we have seen a modest increase in shark attacks in some regions. It is my hope that, perhaps through this book, more and more people realize the beauty, importance, and plight of the white shark and the over 500 species of sharks worldwide. We can and need to coexist with these animals because without them, the health of the ocean and, perhaps, humanity are jeopardized.

White shark Gretel trapped in the shallows.

9
LITERATURE

Everything you've learned in this book comes from research conducted by white shark biologists. I've been directly involved with white shark research for many years, but certainly not long enough or in enough places to know everything. So, a lot of what I've shared with you comes from studies conducted by other scientists in other parts of the world. I didn't have to call, email, or text them. I simply read their published papers. All of this research was peer-reviewed, which means that it was vetted by the scientific community and deemed worthy of publication in a scientific journal. Although some shoddy research findings can slip through, most published studies are considered credible science. It is in every scientist's best interest to publish their findings because they are typically judged by their publication record.

However, not many of these scientific journals are routinely read by the public. That's because these papers are typically laden with statistics and, quite frankly, are generally viewed as boring. Nonetheless, I encourage shark enthusiasts and, particularly, students who want to be shark biologists to check out the scientific literature. The following is a list of the scientific papers that I mined for this book. Although some are off-limits without a subscription, many are in open-access journals and the abstracts are always available, which give you the major findings. If these papers are simply too much to digest, I fully understand. That's why I wrote this book—to translate them for you!

The format of the following citations is straightforward. It starts with the authors' names, followed by the year and title of the study, then the journal or book with the volume and page numbers. Any good search engine will help you find these papers. Enjoy!

Anderson, J.M., Burns, E.S., Meese, E.N., Farrugia, T.J., Stirling, B.S., White, C.F., Logan, R.K., O'Sullivan, J., Winkler, C., and Lowe, C.G. (2021) Interannual nearshore habitat use of young of the year white sharks off Southern California. Frontiers in Marine Science, 8, p.238

Anderson, J.M., Clevenstine, A.J., Stirling, B.S., Burns, E.S., Meese, E.N., White, C.F., Logan, R.K., O'Sullivan, J., Rex, P.T., May, J., and Lyons, K. (2021) Non-random co-occurrence of juvenile white sharks (Carcharodon carcharias) at seasonal aggregation sites in southern California. Frontiers in Marine Science, 8, p.688505

Anderson, J.M., Spurgeon, E., Stirling, B.S., May III, J., Rex, P.T., Hyla, B., McCullough, S., Thompson, M., and Lowe, C.G. (2022) High resolution acoustic telemetry reveals swim speeds and inferred field metabolic rates in juvenile white sharks (*Carcharodon carcharias*). PloS one, 17(6), p.e0268914

Anderson, S.D., Henderson, R P., Pyle, P. and Ainley, D.G. (1996) Observations of white shark reactions to un-baited decoys. In: Great White Sharks: The Biology of *Carcharodon carcharias* (Klimley, A. P., Ainley, D. G., eds), pp. 223–228. Academic Press, San Diego, CA

Anderson, S.D., Klimley, A.P., Pyle, P., and Henderson, R.P. (1996) Tidal height and white shark predation at the Farallon Islands, California. In: Great White Sharks: The Biology of *Carcharodon carcharias* (Klimley, A. P., Ainley, D. G., eds), pp. 275–279. Academic Press, San Diego, CA

Anderson S.D., Pyle, P. (2003) A temporal, sex-specific occurrence pattern among white sharks at the South Farallon Islands, California. California Fish and Game, 89: 96–101

Anderson, S.D., Becker, B.H. and Allen, S.G. (2008) Observations and prey of white sharks, *Carcharodon carcharias*, at Point Reyes National Seashore: 1982–2004. California Fish and Game 94: 33–43

Anderson, S.D., Chapple, T.K., Jorgensen, S.J., Klimley, A.P., and Block, B.A. (2011) Long-term individual identification and site fidelity of white sharks, *Carcharodon carcharias*, off California using dorsal fins. Marine Biology 158(6):1233–1237

Andreotti, S., Von Der Heyden, S., Henriques, R., Rutzen, M., Meÿer, M., Oosthuizen, H., and Matthee, C.A. (2016) New insights into the evolutionary history of white sharks, *Carcharodon carcharias*. Journal of Biogeography, 43: 328–339

Apps, K., Dimmock, K., Lloyd, D. and Huveneers, C. (2016) In the water with white sharks (*Carcharodon carcharias*): participants' beliefs toward cage-diving in Australia. Anthrozoös, 29(2):231-245

Apps, K., Dimmock, K., and Huveneers, C. (2018) Turning wildlife experiences into conservation action: can white shark cage-dive tourism influence conservation behaviour? Marine Policy, 88:108–115

Beaudry, M.C., Hussey, N.E., McMeans, B.C., McLeod, A.M., Wintner, S.P., Cliff, G., Dudley, S.F., and Fisk, A.T. (2015) Comparative organochlorine accumulation in two ecologically similar shark species (*Carcharodon carcharias* and *Carcharhinus obscurus*) with divergent uptake based on different life history. Environmental Toxicology and Chemistry, 34(9):2051-2060

Becerril-García, E.E., Hoyos-Padilla, E.M., Micarelli, P., Galván-Magaña, F. and Sperone, E. (2019) The surface behaviour of white sharks during ecotourism: A baseline for monitoring this threatened species around Guadalupe Island, Mexico. Aquatic Conservation: Marine and Freshwater Ecosystems, 29(5):773-782.

Becerril-García, E.E., Bernot-Simon, D., Arellano-Martínez, M., Galván-Magaña, F., Santana-Morales, O. and Hoyos-Padilla, E.M. (2020) Evidence of interactions between white sharks and large squids in Guadalupe Island, Mexico. Scientific Reports, 10(1):17158.

Benson, J. F., Jorgensen, S. J., O'Sullivan, J. B., Winkler, C., White, C. F., Garcia-Rodriguez, E., et al. (2018) Juvenile survival, competing risks, and spatial variation in mortality risk of a marine apex predator. Journal of Applied Ecology, 55:2888–2897

Bernal, D., Dickson, K.A., Shadwick, R.E., and Graham, J.B. (2001) Analysis of the evolutionary convergence for high performance swimming in lamnid sharks and tunas. Comparative Biochemistry and Physiology, Part A, 129(2–3):695–726

Bigelow, H.B., Schroeder, W.C. (1948) Fishes of the western North Atlantic. Part 1 (lancelets, cyclostomes, sharks). Yale University Press, New Haven, CT

Blower, D.C., Pandolfi, J. M., Bruce, B.D., Gomez-Cabrera, M.D., & Ovenden, J.R. (2012) Population genetics of Australian white sharks reveals fine-scale spatial structure, transoceanic dispersal events and low effective population sizes. Marine Ecology Progress Series, 455:229–244

Boldrocchi, G., Kiszka, J., Purkis, S., Storai, T., Zinzula, L., and Burkholder, D. (2017) Distribution, ecology, and status of the white shark, *Carcharodon carcharias*, in the Mediterranean Sea. Reviews in Fish biology and Fisheries, 27:515-534.

Bonfil, R., Francis, M.P., Duffy, C., Manning, M.J., and O'Brien, S. (2010) Large-scale tropical movements and diving behavior of white sharks *Carcharodon carcharias* tagged off New Zealand. Aquatic Biology, 8:115–123

Bonfil, R., Meÿer, M., Scholl, M.C., Johnson, R., O'Brien, S., Oosthuizen, H., Swanson, S., Kotze, D., and Paterson, M. (2005) Transoceanic migration, spatial dynamics, and population linkages of white sharks. Science, 310:100-103.

Boustany, A.M., Davis, S.F., Pyle, P., Anderson, S.D., Le Boeuf, B.J., and Block, B.A. (2002) Expanded niche for white sharks. Nature, 415: 35–36

Bowlby, H.D., and Gibson, A.J.F. (2020) Implications of life history uncertainty when evaluating status in the Northwest Atlantic population of white shark (*Carcharodon carcharias*). Ecology and Evolution, 10:4990–5000

Bowlby, H. D., Joyce, W. N., Winton, M.V., Coates, P. J., and Skomal, G.B. (2022) Conservation implications of white shark (*Carcharodon carcharias*) behaviour at the northern extent of their range in the Northwest Atlantic. Canadian Journal of Fisheries and Aquatic Sciences, 79(11):1843-1859.

Braccini, M., Taylor, S., Bruce, B., and McAuley, R. (2017) Modelling the population trajectory of West Australian white sharks. Ecological Modelling, 360:363–377

Bradford, R.W., Hobday, A.J., and Bruce, B.D. (2012) Identifying juvenile white shark behavior from electronic tag data. In: Global Perspectives on the Biology and Life History of the White Shark, Domeier, M.L. (ed.). CRC Press, Boca Raton, FL. pp. 255–270.

Braun, C.D., Gaube, P., Sinclair-Taylor, T.H., Skomal, G.B., and Thorrold, S.R. (2019) Mesoscale eddies release pelagic sharks from thermal constraints to foraging in the ocean twilight zone. Proceedings of the National Academy of Sciences, 116:17187–17192

Brena, P.F., Mourier, J., Planes, S., and Clua, E. (2015) Shark and ray provisioning: functional insights into behavioral, ecological and physiological responses across multiple scales. Marine Ecology Progress Series, 538:273–283

Brown, A.C., Lee, D.E., Bradley, R.W., and Anderson, S. (2010) Dynamics of white shark predation on pinnipeds in California: effects of prey abundance. Copeia, 2010(2):232–238

Bruce, B.D. (1992) Preliminary observations on the biology of the white shark, *Carcharodon carcharias*, in South Australian waters. Australian Journal of Marine and Freshwater Research, 43:1–11

Bruce, B.D. (2015) A Review of Cage-Diving Impacts on White Shark Behaviour and Recommendations for Research and the Industry's Management in New Zealand. Hobart: Department of Conservation.

Bruce, B.D., and Bradford, R.W. (2012) Habitat use and spatial dynamics of juvenile white sharks, *Carcharodon carcharias*, in eastern Australia. In: Global Perspectives on the Biology and Life History of the White Shark, Domeier, M.L. (ed.). CRC Press, Boca Raton, FL., pp. 225–254

Bruce, B.D., Stevens, J.D., and Malcolm, H. (2006) Movements and swimming behaviour of white sharks (*Carcharodon carcharias*) in Australian waters. Marine Biology, 150(2):161–172

Bruce, B.D., and Bradford, R.W. (2013) The effects of shark cage-diving operations on the behaviour and movements of white sharks, *Carcharodon carcharias*, at the Neptune Islands, South Australia. Marine Biology, 160:889–907

Bruce, B.D., Stevens, J.D., and Bradford, R.W. (2005) Site Fidelity, Residence Times and Home Range Patterns of White Sharks Around Pinniped Colonies. Hobart: Australian Government Department of Environment and Heritage.

Bruce, B.D., Stevens, J.D., and Malcolm, H. (2006) Movements and swimming behaviour of white sharks (*Carcharodon carcharias*) in Australian waters. Marine. Biology, 150:161–172

Bruce, B., Bradford, R., Bravington, M., Feutry, P., Grewe, P., Gunasekera, R., Harasti, D., Hillary, R. and Patterson, T.A. (2018) A national assessment of the status of white sharks. National Environmental Science Program Marine Biodiversity Hub: Hobart, Australia

Burgess, G.H., and Callahan, M. (1996) Worldwide patterns of white shark attacks on humans. In: Great White Sharks: the Biology of *Carcharodon carcharias*, Klimley, A. P., and Ainley, D. G. (eds.), Academic Press, San Diego, CA, pp. 457–469

Burgess, G.H., Bruce, B.D., Cailliet, G.M., Goldman, K.J., Grubbs, R.D., Lowe, C.G., MacNeil, M.A., Mollet, H.F., Weng, K.C., and O'Sullivan, J.B. (2014) A re-evaluation of the size of the white shark (*Carcharodon carcharias*) population off California, USA. PloS One, 9(6), p.e98078

Butcher, P.A., Piddocke, T.P., Colefax, A.P., Hoade, B., Peddemors, V.M., Borg, L., and Cullis, B.R. (2019) Beach safety: can drones provide a platform for sighting sharks? Wildlife Research 46(8):701–712

Butcher, P.A., Colefax, A.P., Gorkin III, R.A., Kajiura, S.M., López, N.A., Mourier, J., Purcell, C.R., Skomal, G.B., Tucker, J.P., Walsh, A.J., and Williamson, J.E. (2021) The drone revolution of shark science: A review. Drones, 5(1), 8

Cailliet, G.M., Natanson, L.J., Welden, B.A., and Ebert, D.A. (1985) Preliminary studies on the age and growth of the white shark, *Carcharodon carcharias*, using vertebral bands. Memoirs of the Southern California Academy of Sciences, 9:49–60

Carey, F.G., Kanwisher, J.W., Brazier, O., Gabrielson, G., Casey, J.G., and Pratt, H.L. (1982) Temperature and activities of a white shark, *Carcharodon carcharias*. Copeia, 1982:254–260

Carey, F.G., Casey, J.G., Pratt, H.L., Urquhart, D., and McCosker, J.E. (1985) Temperature, heat production, and heat exchange in lamnid sharks. Memoirs of the Southern California Academy of Sciences, 9: 92–108

Carlisle, A.B., Kim, S.L., Semmens, B.X., Madigan, D.J., Jorgensen, S.J., Perle, C.R., Anderson, S.D., Chapple, T.K., Kanive, P.E. and Block, B.A. (2012) Using stable isotope analysis to understand the migration and trophic ecology of northeastern Pacific white sharks (*Carcharodon carcharias*). PloS one, 7(2), p.e30492

Carlson, J.K., Goldman, K.J., and Lowe, C.G. (2004) Metabolism, energetic demand, and endothermy. Biology of sharks and their relatives. 10, 269–286

Casey, J.G., and Pratt, H.L., Jr. (1985) Distribution of the white shark, *Carcharodon carcharias*, in the western North Atlantic. Memoirs of the Southern California Academy of Sciences, 9:2–14

Castro, J.I. (2011) The sharks of North America. Oxford University Press, New York

Castro, J.I. (2012) A summary of observations on the maximum size attained by the white shark, *Carcharodon carcharias*. In: Global Perspectives on the Biology and Life History of the White Shark, Domeier, M.L. (ed.). CRC Press, Boca Raton, FL. pp. 85–90

Chapple, T.K., Jorgensen, S.J., Anderson, S.D., Kanive, P.E., Klimley, A.P., Botsford, L.W., and Block, B.A. (2011) A first estimate of white shark, *Carcharodon carcharias*, abundance off Central California. Biology Letters, 7(4):581-583

Chapple, T.K., Chambert, T., Kanive, P.E., Jorgensen, S.J., Rotella, J.J., Anderson, S.D., Carlisle, A.B. and Block, B.A. (2016) A novel application of multi-event modeling to estimate class segregation in a highly migratory oceanic vertebrate. Ecology, 97(12):3494-3502

Chaprales, W., Lutcavage, M., Brill, R., Chase, B., and Skomal, G. (1998) Harpoon method for attaching ultrasonic and 'popup' satellite tags to giant bluefin tuna and large pelagic fishes. Marine Technology Society Journal, 32:104–105

Chapuis, L., Collin, S.P., Yopak, K.E., McCauley, R.D., Kempster, R.M., Ryan, L.A., Schmidt, C., Kerr, C.C., Gennari, E., Egeberg, C.A., and Hart, N.S. (2019) The effect of underwater sounds on shark behaviour. Scientific reports, 9(1):1-11

Chin, A., Kyne, P.M., Walker, T.I., and McAuley, R.B. (2010) An integrated risk assessment for climate change: analysing the vulnerability of sharks and rays on Australia's Great Barrier Reef. Global Chang Biology, 16:1936–1953

Christiansen, H.M., Lin, V., Tanaka, S., Velikanov, A., Mollet, H.F., Wintner, S.P., Fordham, S.V., Fisk, A.T., and Hussey, N.E. (2014) The last frontier: catch records of white sharks (*Carcharodon carcharias*) in the Northwest Pacific Ocean. PLoS One, 9(4), p.e94407

Christiansen, H.M., Campana, S.E., Fisk, A.T., Cliff, G., Wintner, S.P., Dudley, S.F., Kerr, L.A. and Hussey, N.E., (2016) Using bomb radiocarbon to estimate age and growth of the white shark, *Carcharodon carcharias*, from the southwestern Indian Ocean. Marine Biology, 163:1-13

Cione, A., and Barla, M.J. (2008) Causes and contrasts in current and past distribution of the white shark (Lamniformes: *Carcharodon carcharias*) off southeastern South America. Revista del Museo Argentino de Ciencias Naturales, 10:175–184

Cliff, G., Dudley, S.F.J., and Davis, B. (1989) Sharks caught in the protective gill nets off Natal, South Africa. 2. The great white shark *Carcharodon carcharias* (Linnaeus). South African Journal of Marine Science, 8:131–144

Cliff, G., Van der Elst, R.P., Govender, A., Witthuhn, T.K., and Bullen, E.M. (1996) First estimates of mortality and population size of White Sharks on the South African coast. In: Great White Sharks: the Biology of *Carcharodon carcharias*, Klimley, A.P., and Ainley, D.G. (eds.), Academic Press, San Diego, CA, pp. 393-400

Collier, R. S., Marks, M. and Warner, R.W. (1996) White shark attacks on inanimate objects along the Pacific coast of North America. In: Great White Sharks: the Biology of *Carcharodon carcharias*, Klimley, A.P., and Ainley, D.G. (eds.), Academic Press, San Diego, CA, pp.217–223

Collin, S.P. (2018) Scene through the eyes of an apex predator: a comparative analysis of the shark visual system. Clinical and Experimental Optometry, 101(5):624-640

Compagno, L.J.V. (1991) Government protection for the great white shark (*Carcharodon carcharias*) in South Africa. South African Journal of Marine Science, 87:284–285

Compagno, L.J.V. (2001) Sharks of the world. An annotated and illustrated catalogue of the shark species known to date. Bullhead, mackerel and carpet sharks (Heterodontiformes, Lamniformes and Orectolobiformes), Vol. 2. FAO Species Catalogue for Fisheries Purposes 1, pp.1–269

Cooper, J.A., Pimiento, C., Ferrón, H.G., and Benton, M.J. (2020) Body dimensions of the extinct giant shark *Otodus megalodon*: a 2D reconstruction. Scientific Reports, 10(1), p.14596

Cooper, J.A., Hutchinson, J.R., Bernvi, D.C., Cliff, G., Wilson, R.P., Dicken, M.L., Menzel, J., Wroe, S., Pirlo, J., and Pimiento, C. (2022) The extinct shark *Otodus megalodon* was a transoceanic superpredator: inferences from 3D modeling. Science Advances, 8(33), p.eabm9424

Crossley, R., Collins, C.M., Sutton, S.G., and Huveneers, C. (2014) Public perception and understanding of shark attack mitigation measures in Australia. Human Dimensions of Wildlife, 19:154–165

Curtis, T.H., Bruce, B.D., Cliff, G., Dudley, S.F., Klimley, A.P., Kock, A., Lea, R.N., Lowe, C.G., McCosker, J., Skomal, G., and Werry, J. (2012) Responding to the risk of white shark attack. In: Global Perspectives on the Biology and Life History of the White Shark, Domeier, M.L. (ed.). CRC Press, Boca Raton, FL. pp.477-510

Curtis, T.H., McCandless, C.T., Carlson, J.K., Skomal, G.B., Kohler, N.E., Natanson, L.J., Burgess, G.H., Hoey, J.J., and Pratt Jr, H.L. (2014) Seasonal distribution and historic trends in abundance of white sharks, *Carcharodon carcharias*, in the western North Atlantic Ocean. PloS One, 9(6), p.e99240

Curtis, T.H., Metzger, G., Fischer, C., McBride, B., McCallister, M., Winn, L.J., Quinlan, J., and Ajemian, M.J. (2018) First insights into the movements of young-of-the-year white sharks (*Carcharodon carcharias*) in the western North Atlantic Ocean. Scientific Reports, 8(1):1-8

De Vos, A., Justin, O., Riain, M., Meyer, M. A., Kotze, P. G. H., and Kock, A.A. (2015) Behavior of Cape fur seals (*Arctocephalus pusillus pusillus*) in relation to temporal variation in predation risk by white sharks (*Carcharodon carcharias*) around a seal rookery in False Bay, South Africa. Marine Mammal Science, 31:1118–1131

De Vos, A., Justin, O., Riain, M., Meÿer, M.A., Kotze, P.G.H., and Kock, A.A. (2015) Behavior of Cape fur seals (*Arctocephalus pusillus pusillus*) in response to spatial variation in white shark (*Carcharodon carcharias*) predation risk. Marine Mammal Science, 31:1234–1251

Dewar, H., Domeier, M., and Nasby-Lucas, N. (2004) Insights into young of the year white shark, *Carcharodon carcharias*, behavior in the Southern California Bight. Environmental Biology of Fishes, 70:133–143

Dicken, M.L., and Booth, A.J. (2013) Surveys of white sharks (*Carcharodon carcharias*) off bathing beaches in Algoa Bay, South Africa. Marine and Freshwater Research, 64:530–539

Dickson, K.A., Gregorio, M.O., Gruber, S.J., Loefler, K.L., Tran, M., and Terrell, C. (1993) Biochemical indices of aerobic and anaerobic capacity in muscle tissues of California elasmobranch fishes differing in typical activity level. Marine Biology, 117:185–193

Domeier, M.L. (2009) Experimental scavenging preference for the adult white shark, *Carcharodon carcharias*. California Fish and Game 95:140–145

Domeier, M.L. (2012) A new life-history hypothesis for white sharks, *Carcharodon carcharias*, in the Northeastern Pacific. In: Global Perspectives on the Biology and Life History of the White Shark, Domeier, M.L. (ed.). CRC Press, Boca Raton, FL. pp.199–224

Domeier, M.L. (2012) Global Perspectives on the Biology and Life History of the White Shark. CRC Press, Boca Raton, FL

Domeier, M.L., and Nasby-Lucas, N. (2007) Annual re-sightings of photographically identified white sharks (*Carcharodon carcharias*) at an eastern Pacific aggregation site (Guadalupe Island, Mexico). Marine Biology 150:977–984

Domeier, M.L., and Nasby-Lucas, N. (2008) Migration patterns of white sharks *Carcharodon Carcharias*, tagged at Guadalupe Island, Mexico, and identification of an eastern Pacific shared offshore foraging area. Marine Ecology Progress Series, 370:221–237

Domeier, M.L., and Nasby-Lucas, N. (2012) Sex-specific migration patterns and sexual segregation of adult white sharks, *Carcharodon carcharias*, in the Northeastern Pacific. In: Global Perspectives on the Biology and Life History of the White Shark, Domeier, M.L. (ed.). CRC Press, Boca Raton, FL. pp.133–146

Domeier, M.L., and Nasby-Lucas, N. (2013) Two-year migration of adult female white sharks (*Carcharodon carcharias*) reveals widely separated nursery areas and conservation concerns. Animal Biotelemetry, 1:2

Domeier, M.L., Nasby-Lucas, N., and Lam, C.H. (2012) Fine-scale habitat use by white sharks at Guadalupe Island, Mexico. In: Global Perspectives on the Biology and Life History of the White Shark, Domeier, M.L. (ed.). CRC Press, Boca Raton, FL. pp.121–132

Dudley, S.F.J. (1997) A comparison of the shark control programs of New South Wales and Queensland (Australia) and KwaZulu-Natal (South Africa). Ocean and Coastal Management 34(1):1–27

Dudley, S.F.J., and Cliff, G. (2010) Shark control: methods, efficacy, and ecological impact. In: Sharks and their Relatives II: Biodiversity, Adaptive Physiology, and Conservation, Carrier, J.C., Musick, J.A., and Heithaus, M.R. (eds.). CRC Press: Boca Raton, FL. pp. 567–592

Duffy, C., Francis, M. P., Manning, M.J., and Bonfil, R. (2012) Regional population connectivity, oceanic habitat, and return migration revealed by satellite tagging of white sharks, *Carcharodon carcharias*, at New Zealand aggregation sites. In: Global Perspectives on the Biology and Life History of the White Shark, Domeier, M.L. (ed.). CRC Press, Boca Raton, FL. pp.301–318

Dulvy, N.K., Fowler, S.L., Musick, J.A., Cavanagh, R.D., Kyne, P.M., Harrison, L.R., Carlson, J.K., Davidson, L.N., Fordham, S.V., Francis, M.P., and Pollock, C.M. (2014) Extinction risk and conservation of the world's sharks and rays. elife, 3, p.e00590

Ehret, D.J., Macfadden, B.J., Jones, D.S., Devries, T.J., Foster, D.A., and Salas-Gismondi, R. (2012) Origin of the white shark *Carcharodon* (Lamniformes: Lamnidae) based on recalibration of the Upper Neogene Pisco Formation of Peru. Palaeontology, 55(6):1139-1153

Engelbrecht, T., Kock, A., Waries, S., and O'Riain, M.J. (2017) Shark spotters: successfully reducing spatial overlap between white sharks (*Carcharodon carcharias*) and recreational water users in False Bay, South Africa. PLoS One 12, e0185335

Estrada, J.A., Rice, A.N., Natanson, L.J., and Skomal, G.B. (2006) Use of isotopic analysis of vertebrae in reconstructing ontogenetic feeding ecology in white sharks. Ecology. 87:829–834

Fallows, C., Martin, R.A., and Hammerschlag, N. (2012) Comparisons between white shark-pinniped interactions at Seal Island (South Africa) with other sites in California. In: Global Perspectives on the Biology and Life History of the White Shark, Domeier, M.L. (ed.). CRC Press, Boca Raton, FL. pp.105–117

Fallows, C., Fallows, M., and Hammerschlag, N. (2016) Effects of lunar phase on predator-prey interactions between white shark (*Carcharodon carcharias*) and Cape fur seals (*Arctocephalus pusillus pusillus*). Environmental Biology of Fishes. 99(11):805–812

Fergusson, I.K. (1996) Distribution and autecology of the white shark in the eastern North Atlantic Ocean and the Mediterranean Sea. In: Great White Sharks: the Biology of *Carcharodon carcharias*, Klimley, A.P., and Ainley, D.G. (eds.), Academic Press, San Diego, CA, pp.321-345

Fergusson, I., Compagno, L.J.V., and Marks, M. (2000) Predation by white sharks *Carcharodon carcharias* (Chondrichthyes: Lamnidae) upon chelonians, with new records from the Mediterranean Sea and a first record of the ocean sunfish *Mola mola* (Osteichthyes: Molidae) as stomach contents. Environmental Biology of Fishes, 58:447–453

Fergusson, I., Compagno, L.J.V., and Marks, M. (2009) *Carcharodon carcharias*. The IUCN Red List of Threatened Species 2009: e.T3855A10133872

Ferretti, F., Myers, R.A., Serena, F., and Lotze, H.K. (2008) Loss of large predatory sharks from the Mediterranean Sea. Conservation Biology, 22(4):952–964

Ferretti, F., Worm, B., Britten, G.L., Heithaus, M.R., and Lotze, H. K. (2010) Patterns and ecosystem consequences of shark declines in the ocean. Ecology Letters, 13(8):1055–1071

Ferretti, F., Jorgensen, S., Chapple, T.K., De Leo, G., and Micheli, F. (2015). Reconciling predator conservation with public safety. Frontiers in Ecology and the Environment, 13:412–417

Fisheries and Oceans Canada. (2021) Recovery strategy for the white shark (*Carcharodon carcharias*) in Atlantic Canadian waters. Species at Risk Act recovery strategy series. Fisheries and Oceans Canada, Ottawa, ON.

Francis, M.P. (1996) Observations on a pregnant white shark with a review of reproductive biology. In: Great White Sharks: the Biology of *Carcharodon carcharias*, Klimley, A.P., and Ainley, D.G. (eds.), Academic Press, San Diego, CA, pp.157–172

Francis, M.P., Duffy, C.A.J., Bonfil, R., and Manning, M.J. (2012) The third dimension: vertical habitat use by white sharks, *Carcharodon carcharias*, in New Zealand and in Oceanic and Tropical Waters of the Southwest Pacific Ocean. In: Global Perspectives on the Biology and Life History of the White Shark, Domeier, M.L. (ed.). CRC Press, Boca Raton, FL., pp.319–342

French, G.C.A., Rizzuto, S., Stürup, M., Inger, R., Barker, S., van Wyk, J.H., Towner, A.V., and Hughes, W.O.H. (2018) Sex, size and isotopes: cryptic trophic ecology of an apex predator, the white shark *Carcharodon carcharias*. Marine Biology, 165:1-11

French, G.C.A., Stürup, M., Rizzuto, S., Van Wyk, J.H., Edwards, D., Dolan, R.W., Wintner, S.P., Towner, A.V., and Hughes, W.O.H. (2017) The tooth, the whole tooth and nothing but the tooth: tooth shape and ontogenetic shift dynamics in the white shark *Carcharodon carcharias*. Journal of Fish Biology, 91(4):1032-1047

Gallagher, A.J., and Huveneers, C. (2018) Emerging challenges to shark-diving tourism. Marine Policy, 96:9–12

Goldman, K.J. (1997) Regulation of body temperature in the white shark, *Carcharodon carcharias*. Journal of Comparative Physiology B, 167:423-429

Goldman, K.J., and Anderson, S.D. (1999) Space utilization and swimming depth of white sharks, *Carcharodon carcharias*, at the South Farallon Islands, Central California. Environmental Biology of Fishes, 56:351–364

Goldman, K.J., Anderson, S.D., McCosker, J.E. and Klimley, A.P. (1996) Temperature, swimming depth, and movements of a white shark at the South Farallon Islands, California. In: Great White Sharks: the Biology of *Carcharodon carcharias*, Klimley, A.P., and Ainley, D.G. (eds.), Academic Press, San Diego, CA, pp.111–120

Gooden, A., Clarke, T.M., Meyer, L., and Huveneers, C. (2023) Wildlife tourism has little energetic impact on the world's largest predatory shark. Animal Behaviour (2023).

Gubili, C., Bilgin, R., Kalkan, E., Karhan, S.Ü., Jones, C.S., Sims, D.W., Kabasakal, H., Martin, A.P., and Noble, L.R. (2011) Antipodean white sharks on a Mediterranean walkabout? Historical dispersal leads to genetic discontinuity and an endangered anomalous population. Proceedings of the Royal Society B: Biological Sciences, 278(1712):1679-1686

Gubili, C., Duffy, C.A., Cliff, G., Wintner, S.P., Shivji, M., Chapman, D., Bruce, B.D., Martin, A.P., Sims, D.W., Jones, C.S. and Noble, L.R. (2012) Application of molecular genetics for conservation of the white shark, *Carcharodon carcharias*, L. 1758. In: Global Perspectives on the Biology and Life History of the White Shark, Domeier, M.L. (ed.). CRC Press, Boca Raton, FL, pp.357-380

Gubili, G., Robinson, C.E.C., Cliff, G., Wintner, S.P., de Sabata, E., De Innocentiis, S., and Jones, C.S. (2015) DNA from historical and trophy samples provides insights into white shark population origins and genetic diversity. Endangered Species Research, 27(3):233–241

Gruber, S.H., and Cohen, J.L. (1985) Visual system of the white shark, *Carcharodon carcharias*, with emphasis on retinal structure. Memoirs of the Southern California Academy of Sciences, 9:61–72

Gudger, E.W. (1950) A boy attacked by a shark, July 25, 1936 in Buzzard's Bay, Massachusetts with notes on attacks by another shark along the New Jersey coast in 1916. American Midland Naturalist 44(3):714–719

Hamady, L.L., Natanson, L.J., Skomal, G.B., and Thorrold S.R. (2014) Vertebral bomb radiocarbon suggests extreme longevity in white sharks. PLoS One, 9, e84006

Hammerschlag, N., Martin, R.A., and Fallows, C. (2006) Effects of environmental conditions on predator–prey interactions between white sharks (*Carcharodon carcharias*) and Cape fur seals (*Arctocephalus pusillus pusillus*) at Seal Island, South Africa. Environmental Biology of Fishes 76(2–4):341–350

Hammerschlag, N., Martin, R.A., Fallows, C., Collier, R.S. & Lawrence, R. (2012) Investigatory behaviour toward surface objects and nonconsumptive strikes on seabirds by white sharks, *Carcharodon carcharias*, at Seal Island, South Africa (1997–2010). In: Global Perspectives on the Biology and Life History of the White Shark, Domeier, M.L. (ed.). CRC Press, Boca Raton, FL., pp.91–103

Hara, M., Maharaj, I., and Pithers, L. (2003) Marine-based tourism in Gansbaai: a socio-economic study. Final Report Dept. Environmental Affairs and Tourism, 55

Heupel, M.R., Kanno, S., Martins, A.P., and Simpfendorfer, C.A. (2018) Advances in understanding the roles and benefits of nursery areas for elasmobranch populations. Marine and Freshwater Research, 70(7):897–907

Hewitt, A.M., Kock, A.A., Booth, A.J., and Griffiths, C.L. (2018) Trends in sightings and population structure of white sharks, *Carcharodon carcharias*, at Seal Island, False Bay, South Africa, and the emigration of subadult female sharks approaching maturity. Environmental Biology of Fishes, 101:39–54

Hillary, R.M., Bravington, M.V., Patterson, T.A., Grewe, P., Bradford, R., Feutry, P., Gunasekera, R., Peddemors, V., Werry, J., Francis, M.P. and Duffy, C.A.J. (2018) Genetic relatedness reveals total population size of white sharks in eastern Australia and New Zealand. Scientific reports, 8(1):2661

Hoenicka, M.A.K., Andreotti, S., Carvajal-Chitty, H. and Matthee, C.A. (2022) The role of controlled human-animal interactions in changing the negative perceptions towards white sharks, in a sample of white shark cage diving tours participants. Marine Policy, 143, p.105130

Hoyos-Padilla,M.E. (2009) Movement patterns of the white shark (*Carcharodon carcharias*) at Guadalupe Island, Mexico. PhD Thesis, Centro Interdisciplinario de Ciencias Marinas, Instituto Politécnico Nacional, La Paz, B.C.S., Mexico.

Hoyos-Padilla, E.M., Klimley, A.P., Galván-Magaña, F., and Antoniou, A. (2016) Contrasts in the movements and habitat use of juvenile and adult white sharks (*Carcharodon carcharias*) at Guadalupe Island, Mexico. Animal Biotelemetry, 4:1-14

Hussey, N.E., McCann, H.M., Cliff, G., Dudley, S.F.J., Wintner, S.P., and Fisk, A.T. (2012) Size-based analysis of diet and trophic position of the white shark, *Carcharodon carcharias*, in South African Waters. In: Global Perspectives on the Biology and Life History of the White Shark, Domeier, M.L. (ed.). CRC Press, Boca Raton, FL, pp.27–50

Huveneers, C., Rogers, P.J., Beckmann, C., Semmens, J., Bruce, B., and Seuront, L. (2013) The effects of cage-diving activities on the fine-scale swimming behaviour and space use of white sharks. Marine Biology, 160:2863–2875

Huveneers, C., Rogers, P.J., Semmens, J.M., Beckmann, C., Kock, A.A., Page, B. and Goldsworthy, S.D. (2013) Effects of an electric field on white sharks: in situ testing of an electric deterrent. PloS One, 8(5), p.e62730

Huveneers, C., Holman, D., Robbins, R., Fox, A., Endler, J.A., and Taylor, A.H. (2015) White sharks exploit the sun during predatory approaches. American Naturalist, 185(4):562–570

Huveneers, C., Meekan, M.G., Apps, K., Ferreira, L.C., Pannell, D., and Vianna, G.M.S. (2017) The economic value of shark-diving tourism in Australia. Reviews in Fish Biology and Fisheries, 27:665–680

Huveneers, C., Watanabe, Y.Y., Payne, N.L., and Semmens, J.M. (2018) Interacting with wildlife tourism increases activity of white sharks. Conservation Physiology, 6:coy019

Huveneers, C., Whitmarsh, S., Thiele, M., Meyer, L., Fox, A., and Bradshaw, C.J.A. (2018) Effectiveness of five personal shark-bite deterrents for surfers. PeerJ, 6:e5554

Huveneers, C., Apps, K., Becerril-Garcı´a, E.E., Bruce, B., Butcher, P.A., Carlisle, A.B., Chapple, T.K., Christiansen, H.M., Cliff, G., Curtis, T.H., Daly-Engel, T.S., Dewar, H., Dicken, M.L., Domeier, M.L., Duffy, C.A. J., Ford, R., Francis, M.P., French, G.C.A., Galva´n- Magan˜a, F., Garcı´a-Rodrı´guez, E., Gennari, E., Graham, B., Hayden, B., Hoyos-Padilla, E. M., Hussey, N.E., Jewell, O.J.D., Jorgensen, S.J., Kock, A.A., Lowe, C.G., Lyons, K., Meyer, L., Oelofse, G., On¯ate-Gonza´lez, E.C., Oosthuizen, H., O'Sullivan, J.B., Ramm, K., Skomal, G., Sloan, S., Smale, M.J., Sosa-Nishizaki, O., Sperone, E., Tamburin, E., Towner, A. V., Wcisel, M.A., Weng, K.C., and Werry, J.M. (2018) Future research directions on the 'Elusive' white shark. Frontiers in Marine Science, 5:455

Huveneers, C., Watanabe, Y.Y., Payne, N.L., and Semmens, J.M. (2018) Interacting with wildlife tourism increases activity of white sharks. Conservation Physiology, 6(1), p.coy019.

Huveneers, C., Blount, C., Bradshaw, C.J., Butcher, P.A., Smith, M.P.L., Macbeth, W.G., McPhee, D.P., Moltschaniwskyj, N., Peddemors, V.M., and Green, M. (2024) Shifts in the incidence of shark bites and efficacy of beach-focused mitigation in Australia. Marine Pollution Bulletin, 198, p.115855.

Jaime-Rivera, M., Caraveo-Patiño, J., Hoyos-Padilla, M., and Galván-Magaña, F. (2014) Feeding and migration habits of white shark Carcharodon carcharias (Lamniformes: Lamnidae) from Isla Guadalupe inferred by analysis of stable isotopes δ15N and δ13C. Revista de biologia tropical, 62(2):637-647

Jewell, O.J., Wcisel, M.A., Gennari, E., Towner, A.V., Bester, M.N., Johnson, R.L. and Singh, S. (2011) Effects of smart position only (SPOT) tag deployment on white sharks Carcharodon carcharias in South Africa. PLoS One, 6(11), p.e27242

Jewell, O.J., Johnson, R.L., Gennari, E., and Bester, M.N. (2013) Fine scale movements and activity areas of white sharks (Carcharodon carcharias) in Mossel Bay, South Africa. Environmental Biology of Fishes, 96(7):881–894

Johnson, R., Bester, M.N., Dudley, S.F., Oosthuizen, W.H., Meyer, M., Hancke, L., and Gennari, E. (2009) Coastal swimming patterns of white sharks (Carcharodon carcharias) at Mossel Bay, South Africa. Environmental Biology of Fishes 85(3):189–200

Jorgensen, S.J., Reeb, C.A., Chapple, T.K., Anderson, S., Perle, C., Van Sommeran, S.R., Fritz-Cope, C., Brown, A.C., Klirnley, A P. and Block, B.A. (2010) Philopatry and migration of Pacific white sharks. Proceedings of the Royal Society B, 277:679–688

Jorgensen, S.J., Chapple, T.K., Anderson, S., Hoyos, M., Reeb, C., and Block, B.A. (2012) Connectivity among white shark coastal aggregation areas in the Northeastern Pacific. In: Global Perspectives on the Biology and Life History of the White Shark, Domeier, M.L. (ed.). CRC Press, Boca Raton, FL, pp.159–168

Jorgensen, S.J., Arnoldi, N.S., Estess, E.E., Chapple, T.K., Rückert, M., Anderson, S.D. and Block, B.A. (2012) Eating or meeting? Cluster analysis reveals intricacies of white shark (*Carcharodon carcharias*) migration and offshore behavior. PloS One, 7(10), p.e47819

Jorgensen, S.J., Gleiss, A.C., Kanive, P.E., Chapple, T.K., Anderson, S.D., Ezcurra, J.M., Brandt, W.T. and Block, B.A. (2015) In the belly of the beast: resolving stomach tag data to link temperature, acceleration and feeding in white sharks (*Carcharodon carcharias*). Animal Biotelemetry, 3:1-10

Jorgensen, S.J., Anderson, S., Ferretti, F., Tietz, J.R., Chapple, T., Kanive, P., Bradley, R.W., Moxley, J.H. and Block, B.A. (2019) Killer whales redistribute white shark foraging pressure on seals. Scientific Reports, 9(1), p.6153

Kabasakal, H. (2014) The status of the great white shark (*Carcharodon carcharias*) in Turkey's waters. Mar. Biodiversity Records, 7:1–8

Kabasakal, H., and Gedikoğlu, Ö.S. (2008) Two new-born great white sharks, *Carcharodon carcharias* (Linnaeus, 1758) (Lamniformes; Lamnidae) from Turkish waters of the north Aegean Sea. Acta Adriatica, 49:125–135

Kanive, P.E., Rotella, J.J., Chapple, T.K., Anderson, S.D., White, T.D., Block, B.A. and Jorgensen, S.J. (2021) Estimates of regional annual abundance and population growth rates of white sharks off central California. Biological Conservation, 257, p.109104

Kempster, R.M., Egeberg, C.A., Hart, N.S., Ryan, L., Chapuis, L., Kerr, C.C., Schmidt, C., Huveneers, C., Gennari, E., Yopak, K.E., and Meeuwig, J.J. (2016) How close is too close? The effect of a non-lethal electric shark deterrent on white shark behaviour. PLoS One, 11(7), p.e0157717

Kim, S. L., Tinker, M.T., Estes, J.A., and Koch, P.L. (2012) Ontogenetic and among-individual variation in foraging strategies of northeast Pacific white sharks based on stable isotope analysis. PLoS One, 7, e45068

Kitchell, J.F., Essington, T.E., Boggs, C.H., Schindler, D.E., and Walters, C.J. (2002) The role of sharks and longline fisheries in a pelagic ecosystem of the central Pacific. Ecosystems, 5:202-216

Klimley, A.P. (1985) The areal distribution and autoecology of the white shark, *Carcharodon carcharias*, off the West Coast of North America. Memoirs of the Southern California Academy of Sciences, 9:15–40

Klimley, A.P. (1994) The predatory behaviour of the white shark. American Scientist, 82:122–133

Klimley, A.P., Anderson, S.D., Pyle, P., and Henderson, R.P. (1992) Spatiotemporal patterns of white shark (*Carcharodon carcharias*) predation at the South Farallon Islands, California. Copeia, 1992(3):680–690

Klimley, A.P., and Ainley, D.G. (1996) Great white sharks: the biology of *Carcharodon carcharias*. Academic Press, San Diego, CA

Klimley, A.P., Pyle, P., and Anderson, S.D. (1996) The behaviour of white sharks and their pinniped prey during predatory attacks. In: Great White Sharks: the Biology of *Carcharodon carcharias*, Klimley, A.P., and Ainley, D.G. (eds.), Academic Press, San Diego, CA, pp.175–191.

Klimley, A.P., Pyle, P., and Anderson, S. D. (1996) Tail slap and breach: agonistic displays among white sharks. In: Great White Sharks: the Biology of *Carcharodon carcharias*, Klimley, A.P., and Ainley, D.G. (eds.), Academic Press, San Diego, CA, pp.241–255

Klimley, A.P., Le Boeuf, B.J., Cantara, K.M., Richert, J.E., Davis, S.F., Sommeran, S.V., and Kelly, J. T. (2001) The hunting strategy of white sharks (*Carcharodon carcharias*) near a seal colony. Marine Biology, 138:617–636

Kock, A., and Johnson, R.L. (2006) White shark abundance: not a causative factor in numbers of shark bite incidents. In: Finding a Balance: White Shark Conservation and Recreational Safety in the Inshore Waters of Cape Town, South Africa, Nel, D.C., and Peschak, T.P. (eds.), Cape Town: WWF South Africa, pp.1–19

Kock, A., Titley, S., Petersen, W., Sikweyiya, M., Tsotsobe, S., Colenbrander, D., Gold, H., and Oelofse, G. (2012) Shark Spotters: a pioneering shark safety program in Cape Town, South Africa. In: Global Perspectives on the Biology and Life History of the White Shark, Domeier, M.L. (ed.). CRC Press, Boca Raton, FL, pp.447–466

Kock, A., O'Riain, M. J., Mauff, K., Mey¨er, M., Kotze, D., and Griffiths, C. (2013) Residency, habitat use and sexual segregation of white sharks, *Carcharodon carcharias* in False Bay, South Africa. PLoS One, 8(1), e55048

Kock, A.A., Photopoulou, T., Durbach, I., Mauff, K., Mey¨er, M., Kotze, D., Griffiths, C.L., and O'Riain, M.J. (2018) Summer at the beach: spatio-temporal patterns of white shark occurrence along the inshore areas of False Bay, South Africa. Movement Ecology, 6:7

Kohler, N.E. and Turner, P.A. (2019). Distributions and movements of Atlantic shark species: a 52-year retrospective atlas of mark and recapture data. Marine Fisheries Review, 81(2):1-93

Lafferty, K.D., Benesh, K.C., Mahon, A.R., Jerde, C.L., and Lowe, C.G. (2018) Detecting southern California's white sharks with environmental DNA. Frontiers in Marine Science, 5:355

Laroche, K.R., Kock, A.A., Dill, L.M., and Oosthuizen, H. (2007) Effects of provisioning ecotourism activity on the behaviour of white sharks *Carcharodon carcharias*. Marine Ecology Progress Series, 338:199–209

Laroche, R.K., Kock, A.A., Dill, L.M. and Oosthuizen, W.H. (2008) Running the gauntlet: a predator–prey game between sharks and two age classes of seals. Animal Behaviour, 76:1901–1917

Larson, S.E., Daly-Engel, T.S., and Phillips, N.M. (2017) Review of current conservation genetic analyses of Northeast Pacific sharks. Advances in Marine Biology, 77:79–110

Lee, K.A., Butcher, P.A., Harcourt, R.G., Patterson, T.A., Peddemors, V.M., Roughan, M., Harasti, D., Smoothey, A.F., and Bradford, R.W. (2021) Oceanographic conditions associated with white shark (*Carcharodon carcharias*) habitat use along eastern Australia. Marine Ecology Progress Series, 659, pp.143-159.

Leone, A., Puncher, G.N., Ferretti, F., Sperone, E., Tripepi, S., Micarelli, P., Gambarelli, A., Sarà, M., Arculeo, M., Doria, G., and Garibaldi, F. (2020) Pliocene colonization of the Mediterranean by great white shark inferred from fossil records, historical jaws, phylogeographic and divergence time analyses. Journal of Biogeography, 47(5):1119-1129

Levine, M. (1996) Unprovoked attacks by white sharks off the South African coast. In: Great White Sharks: the Biology of *Carcharodon carcharias*, Klimley, A.P., and Ainley, D.G. (eds.), Academic Press, San Diego, CA, pp.435–448

Lingham-Soliar, T. (2005) Dorsal fin in the white shark, *Carcharodon carcharias*: a dynamic stabilizer for fast swimming. Journal of Morphology, 263(1):1-11

Lowe, C.G., Blasius, M.E., Jarvis, E.T., Mason, T.J., Goodmanlowe, G.D., and O'Sullivan, J.B. (2012) Historic fishery interactions with white sharks in the Southern California Bight. In: Global Perspectives on the Biology and Life History of the White Shark, Domeier, M.L. (ed.). CRC Press, Boca Raton, FL, pp.169–186

Lyons, K., Jarvis, E.T., Jorgensen, S.J., Weng, K., O'Sullivan, J., Winkler, C. and Lowe, C.G. (2013) The degree and result of gillnet fishery interactions with juvenile white sharks in southern California assessed by fishery-independent and-dependent methods. Fisheries Research, 147:370-380

Marra, N.J., Stanhope, M.J., Jue, N.K., Wang, M., Sun, Q., Pavinski Bitar, P., Richards, V.P., Komissarov, A., Rayko, M., Kliver, S., and Stanhope, B.J. (2019) White shark genome reveals ancient elasmobranch adaptations associated with wound healing and the maintenance of genome stability. Proceedings of the National Academy of Sciences, 116(10):4446-4455

Martin, R.A. (2007) A review of shark agonistic displays: comparison of display features and implications for shark–human interactions. Marine and Freshwater Behaviour and Physiology, 40:3–34

Martin, R.A., and Hammerschlag, N. (2012) Marine predator–prey contests: ambush and speed versus vigilance and agility. Marine Biology Research, 8(1):90–94

Martin, R.A., Rossmo, D.K., and Hammerschlag, N. (2009) Hunting patterns and geographic profiling of white shark predation. Journal of Zoology, 279:111–118

Martin, R.A., Hammerschlag, N., Collier, R.S., and Fallows, C. (2005) Predatory behaviour of white sharks (*Carcharodon carcharias*) at Seal Island, South Africa. Journal of the Marine Biological Association of the United Kingdom, 85(5):1121–1135

McCosker, J.E. (1985) White shark attack behaviour: observations of and speculations about predator and prey strategies. Memoirs of the Southern California Academy of Sciences, 9:123-135

McCosker, J.E., and Lea, R.N. (1996) White shark attacks in the eastern Pacific Ocean: an update and analysis. In: Great White Sharks: the Biology of *Carcharodon carcharias*, Klimley, A.P., and Ainley, D.G. (eds.), Academic Press, San Diego, CA, pp.419–434

McPherson, J.M., and Myers, R.A. (2009) How to infer population trends in sparse data: examples with opportunistic sighting records for great white sharks. Diversity and Distributions, 15:880–890

Meeuwig, J.J., and Ferreira, L.C. (2014) Moving beyond lethal programs for shark hazard mitigation. Animal Conservation, 17:297–298

Meza-Arce, M.I., Malpica-Cruz, L., Hoyos-Padilla, M.E., Mojica, F.J., Arredondo-García, M.C., Leyva, C., Zertuche-Chanes, R., and Santana-Morales, O. (2020) Unraveling the white shark observation tourism at Guadalupe Island, Mexico: Actors, needs and sustainability. Marine Policy, 119, p.104056

Mickle, M.F. and Higgs, D.M. (2022) Towards a new understanding of elasmobranch hearing. Marine Biology, 169(1):12

Midway, S.R., Wagner, T., and Burgess, G.H. (2019) Trends in global shark attacks. PLoS One, 14(2), e0211049

Morey, G., Martínez, M., Massutí, E., and Moranta, J. (2003) The occurrence of white sharks, *Carcharodon carcharias*, around the Balearic Islands (western Mediterranean Sea). Environmental Biology of Fishes, 68:425–432

Moro, S., Jona-Lasinio, G., Block, B., Micheli, F., De Leo, G., Serena, F., Bottaro, M., Scacco, U., and Ferretti, F. (2020) Abundance and distribution of the white shark in the Mediterranean Sea. Fish and Fisheries, 21(2):338-349

Moxley, J.H., Skomal, G., Chisholm, J., Halpin, P., and Johnston, D. W. (2020) Daily and seasonal movements of Cape Cod gray seals vary with predation risk. Marine Ecology Progress Series, 644:215–228

Mull, C.G., Lyons, K., Blasius, M.E., Winkler, C., O'Sullivan, J.B., and Lowe, C.G. (2013) Evidence of maternal offloading of organic contaminants in white sharks (*Carcharodon carcharias*). PLoS One, 8, e62886

Nasby-Lucas, N., Dewar, H., Lam, C. H., Goldman, K. J., and Domeier, M. L. (2009) White shark offshore habitat: a behavioral and environmental characterization of the eastern Pacific shared offshore foraging area. PLoS One, 4(12), e8163

Natanson, L.J., and Skomal, G.B. (2015) Age and growth of the white shark, *Carcharodon carcharias*, in the western North Atlantic Ocean. Marine and Freshwater Research, 66: 387–398

Natanson, L.J., Skomal, G.B., Hoffmann, S.L., Porter, M.E., Goldman, K.J., and Serra, D. (2018) Age and growth of sharks: do vertebral band pairs record age? Marine and Freshwater Research, 69(9):1440-1452

Nazimi, L., Robbins, W.D., Schilds, A., and Huveneers, C. (2018) Comparison of industry-based data to monitor white shark cage-dive tourism. Tourism Management, 66:263-273

O'Connell, C.P., Andreotti, S., Rutzen, M., Meyer, M., and Matthee, C.A. (2018) Testing the exclusion capabilities and durability of the Sharksafe Barrier to determine its viability as an eco-friendly alternative to current shark culling methodologies. Aquatic Conservation in Marine and Freshwater Ecosystems, 28:252–258

O'Leary, S.J., Feldheim, K.A., and Chapman, D.D. (2013) Novel microsatellite loci for white, *Carcharodon carcharias* and sand tiger sharks, *Carcharias taurus* (order Lamniformes). Conservation. Genetics Resources, 5:627–629

O'Leary, S.J., Feldheim, K.A., Fields, A.T., Natanson, L.J., Wintner, S., Hussey, N., Shivji, M.S., and Chapman, D.D. (2015) Genetic diversity of white sharks, *Carcharodon carcharias*, in the Northwest Atlantic and southern Africa. Journal of Heredity, 106(3):258-265

Oñate-González, E.C., Rocha-Olivares, A., Saavedra-Sotelo, N.C., and Sosa-Nishizaki, O. (2015) Mitochondrial genetic structure and matrilineal origin of white sharks, *Carcharodon carcharias*, in the Northeastern Pacific: implications for their conservation. Journal of Heredity, 106:347–354

Oñate-González, E.C., Sosa-Nishizaki, O., Herzka, S.Z., Lowe, C.G., Lyons, K., Santana-Morales, O., Sepulveda, C., Guerrero-Ávila, C., García-Rodríguez, E., and O'Sullivan, J.B. (2017) Importance of Bahia Sebastian Vizcaino as a nursery area for white sharks (*Carcharodon carcharias*) in the Northeastern Pacific: a fishery dependent analysis. Fisheries Research, 188:125-137

Papastamatiou, Y.P., Mourier, J., TinHan, T., Luongo, S., Hosoki, S., Santana-Morales, O., and Hoyos-Padilla, M. (2022) Social dynamics and individual hunting tactics of white sharks revealed by biologging. Biology Letters, 18(3), p.20210599

Pardini, A.T., Jones, C.S., Noble, L.R., Kreiser, B., Malcolm, H., Bruce, B.D., Stevens, J.D., Cliff, G., Scholl, M.C., Francis, M., and Duffy, C.A. (2001) Sex-biased dispersal of great white sharks. Nature, 412(6843):139-140

Pepin-Neff, C., and Wynter, T. (2018) Shark bites and shark conservation: an analysis of human attitudes following shark bite incidents in two locations in Australia. Conservation Letters, 11:e12407

Pimiento, C., and Clements, C.F. (2014) When did *Carcharocles megalodon* become extinct? A new analysis of the fossil record. PLoS One, 9(10), e111086

Pratt, H.L., Casey, J.G., Conklin, R.B. (1982) Observations on large white sharks, *Carcharodon carcharias*, off Long Island, New York. Fishery Bulletin, 80:153–156

Pratt, H.L. (1996) Reproduction in the male white shark. In: Great White Sharks: the Biology of *Carcharodon carcharias*, Klimley, A.P., and Ainley, D.G. (eds.), Academic Press, San Diego, CA, pp.31-138

Pyle, P., Anderson, S.D., and Ainley, D.G. (1996) Trends in white shark predation at the South Farallon Islands, 1968–1993. In: Great White Sharks: the Biology of *Carcharodon carcharias*, Klimley, A.P., and Ainley, D.G. (eds.), Academic Press, San Diego, CA, pp.375–379

Pyle, P., Klimley, A.P., Anderson, S.D., and Henderson, R.P. (1996) Environmental factors affecting the occurrence and behavior of white sharks at the Farallon Islands, California. In: Great White Sharks: the Biology of *Carcharodon carcharias*, Klimley, A.P., and Ainley, D.G. (eds.), Academic Press, San Diego, CA, pp.281–291

Queiroz, N., Humphries, N.E., Mucientes, G., Hammerschlag, N., Lima, F.P., Scales, K.L., Miller, P.I., Sousa, L.L., Seabra, R., and Sims, D.W. (2016) Ocean-wide tracking of pelagic sharks reveals extent of overlap with longline fishing hotspots. Proceedings of the National Academy of Sciences, 113(6):1582-1587

Queiroz, N., Humphries, N.E., Couto, A., Vedor, M., Da Costa, I., Sequeira, A.M., Mucientes, G., Santos, A.M., et al. (2019) Global spatial risk assessment of sharks under the footprint of fisheries. Nature, 572(7770):461-466

Reid, D.D., Robbins, W.D., and Peddemors, V.M. (2011) Decadal trends in shark catches and effort from the New South Wales, Australia, Shark Meshing Program 1950–2010. Marine and Freshwater Research, 62:676–693

Rex, P.T., May III, J.H., Pierce, E.K., and Lowe, C.G. (2023) Patterns of overlapping habitat use of juvenile white shark and human recreational water users along southern California beaches. Plos one, 18(6), p.e0286575

Rizzari, J.R., Semmens, J.M., Fox, A., and Huveneers, C. (2017) Observations of marine wildlife tourism effects on a non-focal species. Journal of Fish Biology, 91:981–988

Saídi, B., Bradai, M.N., Bouain, A., Guelorget, O., and Capape, C. (2005) Capture of a pregnant female white shark, *Carcharodon carcharias* (Lamnidae) in the Gulf of Gabes (southern Tunisia, central Mediterranean) with comments on oophagy in sharks. Cybium, 29:303–307

Santana-Morales, O., Sosa-Nishizaki, O., Escobedo-Olvera, M.A., OñateGonzález, E.C., O'Sullivan, J.B., and Cartamil, D. (2012) Incidental catch and ecological observations of juvenile white sharks, *Carcharodon carcharias*, in Western Baja California, Mexico: conservation implications. In: Global Perspectives on the Biology and Life History of the White Shark, Domeier, M.L. (ed.). CRC Press, Boca Raton, FL, pp.187–198

Santana-Morales, O., Abadía-Cardoso, A., Hoyos-Padilla, M., Naylor, G.J., Corrigan, S., Malpica-Cruz, L., Aquino-Baleytó, M., Beas-Luna, R., Sepúlveda, C.A., and Castillo-Géniz, J.L. (2020) The smallest known free-living white shark *Carcharodon carcharias* (Lamniformes: Lamnidae): ecological and management implications. Copeia, 108(1):pp.39-46

Santana-Morales, O., Hoyos-Padilla, E.M., Medellín-Ortíz, A., Sepulveda, C., Beas-Luna, R., Aquino-Baleytó, M., Becerril-García, E.E., Arellano-Millán, D., Malpica-Cruz, L., Lorda, J., and Castillo-Géniz, J.L. (2021) How much is too much? A carrying capacity study of white shark cage diving in Guadalupe Island, Mexico. Marine Policy, 131, p.104588

Sato, K., Nakamura, M., Tomita, T., Toda, M., Miyamoto, K., and Nozu, R. (2016) How great white sharks nourish their embryos to a large size: evidence of lipid histotrophy in lamnoid shark reproduction. Biology Open, 5:1211–1215

Semmens, J.M., Payne, N.L., Huveneers, C., Sims, D.W., and Bruce, B.D. (2013) Feeding requirements of white sharks may be higher than originally thought. Scientific Reports, 3:1471

Semmens, J.M., Kock, A.A., Watanabe, Y.Y., Shepard, C.M., Berkenpas, E., Stehfest, K.M., Barnett, A., and Payne, N.L. (2019) Preparing to launch: biologging reveals the dynamics of white shark breaching behaviour. Marine Biology, 166:1-9

Shaw, R.L., Curtis, T.H., Metzger, G., McCallister, M.P., Newton, A., Fischer, G.C., and Ajemian, M.J. (2021) Three-dimensional movements and habitat selection of young white sharks (*Carcharodon carcharias*) across a temperate continental shelf ecosystem. Frontiers in Marine Science, 8:643831

Shimada, K. (2021) The size of the megatooth shark, *Otodus megalodon* (Lamniformes: Otodontidae), revisited. Historical Biology, 33(7):904-911

Shimada, K., Becker, M.A. and Griffiths, M.L. (2021) Body, jaw, and dentition lengths of macrophagous lamniform sharks, and body size evolution in Lamniformes with special reference to 'off-the-scale' gigantism of the megatooth shark, *Otodus megalodon*. Historical Biology, 33(11):2543-2559

Shimada, K., Chandler, R.E., Lam, O.L.T., Tanaka, T., and Ward, D.J. (2017) A new elusive otodontid shark (Lamniformes: Otodontidae) from the lower Miocene, and comments on the taxonomy of otodontid genera, including the 'megatoothed' clade. Historical Biology, 29(5):704-714

Skomal, G.B., Lobel, P.S. and Marshall, G. (2007) The use of animal-borne imaging to assess post-release behaviour as it relates to capture stress in grey reef sharks, *Carcharhinus amblyrhynchos*. Marine Technology Society Journal, 41:44–48

Skomal, G.B., Zeeman, S.I., Chisholm, J.H., Summers, E.L., Walsh, H.J., McMahon, K.W., and Thorrold, S. R. (2009) Transequatorial migrations by basking sharks in the western Atlantic Ocean. Current Biology, 19:1–4

Skomal, G.B., Chisholm, J., and Correia, S. (2012) Implications of increasing pinniped populations on the diet and abundance of white sharks off the coast of Massachusetts. In: Global Perspectives on the Biology and Life History of the White Shark, Domeier, M.L. (ed.). CRC Press, Boca Raton, FL, pp.405–418

Skomal, G.B., Braun, C.D., Chisholm, J.H., and Thorrold, S.R. (2017) Movements of the white shark *Carcharodon carcharias* in the North Atlantic Ocean. Marine Ecology Progress Series, 580:1–16

Skomal, G.B., Hoyos-Padilla, E.M., Kukulya, A.L., and Stokey, R.P. (2015) Subsurface observations of white shark predatory behaviour using an autonomous underwater vehicle. Journal of Fish Biology, 87:1293–1312

Skubel, R.A., Kirtman, B.P., Fallows, C., and Hammerschlag, N. (2018) Patterns of long-term climate variability and predation rates by a marine apex predator, the white shark *Carcharodon carcharias*. Marine Ecology Progress Series, 587:129–139

Soldo, A., and Jardas, I. (2002) Occurrence of great white shark, *Carcharodon carcharias* (Linnaeus, 1758) and basking shark, *Cetorhinus maximus* (Gunnerus, 1758) in the Eastern Adriatic and their protection. Periodicum Biologorum, 104(2):195-202

Sosa-Nishizaki, O., Morales-Bojòìrquez, E., Nasby-Lucas, N., Onþate-Gonzaìlez, E.C., and Domeier, M.L. (2012) Problems with photo identification as a method of estimating abundance of white sharks (*Carcharodon carcharias*): an example from Guadalupe Island, Mexico. In: Global Perspectives on the Biology and Life History of the White Shark, Domeier, M.L. (ed.). CRC Press, Boca Raton, FL, pp.393–404

Spaet, J.L., Butcher, P.A., Manica, A., and Lam, C.H. (2022) Spatial dynamics and fine-scale vertical behaviour of immature eastern Australasian white sharks (*Carcharodon carcharias*). Biology, 11(12):p.1689.

Sperone, E., Micarelli, P., Andreotti, S., Brandmayr, P., Bernabò, I., Brunelli, E., and Tripepi, S. (2012) Surface behaviour of bait-attracted white sharks at Dyer Island (South Africa). Marine Biology Research, 8(10):982-991

Sperone, E., Parise, G., Leone, A., Milazzo, C., Circosta, V., Santoro, G., Paolillo, G., Micarelli, P., and Tripepi, S. (2012) Spatiotemporal patterns of distribution of large predatory sharks in Calabria (central Mediterranean, southern Italy). Acta Adriatica, 53(1):13-24

Spurgeon, E., Anderson, J.M., Liu, Y., Barajas, V.L., and Lowe, C.G. (2022) Quantifying thermal cues that initiate mass emigrations in juvenile white sharks. Scientific Reports, 12(1), p.19874.

Strong, W.R. (1996) Shape discrimination and visual predatory tactics in white sharks. In: Great White Sharks: the Biology of *Carcharodon carcharias*, Klimley, A.P., and Ainley, D.G. (eds.), Academic Press, San Diego, CA, pp.229–240

Strong, W.R., Murphy, R.C., Bruce, B.D., and Nelson, D.R. (1992) Movements and associated observations of bait-attracted white sharks, *Carcharodon carcharias*: a preliminary report. Marine and Freshwater Research, 43(1):13-20

Strong, W.R., Bruce, B.D., Nelson, D.R., and Murphy, R.D. (1996) Population dynamics of white sharks in Spencer Gulf, South Australia. In: Great White Sharks: the Biology of *Carcharodon carcharias*, Klimley, A.P., and Ainley, D.G. (eds.), Academic Press, San Diego, CA, pp.401–414

Tamburin, E., Hoyos-Padilla, M., Sánchez-González, A., Hernández-Herrera, A., Elorriaga-Verplancken, F.R., and Galván-Magaña, F. (2019) New nursery area for white sharks (*Carcharodon carcharias*) in the Eastern Pacific Ocean. Turkish Journal of Fisheries and Aquatic Sciences, 20(4):325-329

Tanaka, K.R., Van Houtan, K.S., Mailander, E., Dias, B.S., Galginaitis, C., O'Sullivan, J., Lowe, C.G., and Jorgensen, S.J. (2021) North Pacific warming shifts the juvenile range of a marine apex predator. Scientific Reports, 11(1):p.3373.

Tanaka, S., Kitamura, T., Mochizuki, T., and Kofuji, K. (2011) Age, growth and genetic status of the white shark (*Carcharodon carcharias*) from Kashima-nada, Japan. Marine and Freshwater Research, 62:548–556

Taylor, J.K.D., Mandelman, J.W., McLellan, W.A., Moore, M.J., Skomal, G.B., Rotstein, D.S., and Kraus, S.D. (2013) Shark predation on North Atlantic right whales (*Eubalaena glacialis*) in the southeastern United States calving ground. Marine Mammal Science, 29:204–212

Templeman, W. (1963) Distribution of sharks in the Canadian Atlantic (with special reference to Newfoundland waters). Bulletin of the Fisheries Research Board of Canada, 140:1–77

Towner, A.V., Wcisel, M.A., Reisinger, R.R., Edwards, D., and Jewell, O.J.D. (2013) Gauging the threat: the first population estimate for white sharks in South Africa using photo identification and automated software. PLoS One, 8, e66035

Towner, A.V., Leos-Barajas, V., Langrock, R., Schick, R.S., Smale, M.J., Kaschke, T., Jewell, O.J., and Papastamatiou, Y.P. (2016) Sex-specific and individual preferences for hunting strategies in white sharks. Functional Ecology, 30(8):1397–1407

Towner, A.V., Watson, R.G.A., Kock, A.A., Papastamatiou, Y., Sturup, M., Gennari, E., Baker, K., Booth, T., Dicken, M., Chivell, W., and Elwen, S. (2022) Fear at the top: killer whale predation drives white shark absence at South Africa's largest aggregation site. African Journal of Marine Science, 44(2):139-152

Towner, A.V., Kock, A.A., Stopforth, C., Hurwitz, D. and Elwen, S.H. (2023) Direct observation of killer whales preying on white sharks and evidence of a flight response. Ecology, 104(1).

Tricas, T.C. (1985) Feeding ethology of the white shark, *Carcharodon carcharias*. Memoirs of the Southern California Academy of Sciences, 9:81–91

Tricas, T.C., and McCosker, J.E. (1984) Predatory behaviour of the white shark (*Carcharodon carcharias*), with notes on its biology. Proceedings of the California Academy of Sciences, 43: 221–238

Verkamp, H.J., Skomal, G., Winton, M., and Sulikowski, J.A. (2021) Using reproductive hormone concentrations from the muscle of white sharks *Carcharodon carcharias* to evaluate reproductive status in the Northwest Atlantic Ocean. Endangered Species Research, 44:231-236

Watanabe, Y.Y., Payne, N.L., Semmens, J.M., Fox, A., and Huveneers, C. (2019) Swimming strategies and energetics of endothermic white sharks during foraging. Journal of Experimental Biology 222(4):jeb185603

Watanabe, Y.Y., Payne, N.L., Semmens, J.M., Fox, A., and Huveneers, C. (2019) Hunting behaviour of white sharks recorded by animal-borne accelerometers and cameras. Marine Ecology Progress Series, 621:221-227

Weltz, K., Kock, A.A., Winker, H., Attwood, C., and Sikweyiya, M. (2013) The influence of environmental variables on the presence of white sharks, *Carcharodon carcharias*, at two popular Cape Town bathing beaches: a generalized additive mixed model. PLoS One, 8(7), e68554

Weng, K.C., Boustany, A.M., Pyle, P., Anderson, S.D., Brown, A., and Block, B.A. (2007) Migration and habitat of white sharks (*Carcharodon carcharias*) in the eastern Pacific Ocean. Marine Biology, 152(4):877–894

Weng, K.C., O'Sullivan, J.B., Lowe, C.G., Winkler, C.E., Dewar, H., and Block, B.A. (2007) Movements, behavior and habitat preferences of juvenile white sharks *Carcharodon carcharias* in the eastern Pacific. Marine Ecology Progress Series, 338:211–224

Werry, J.M., Bruce, B.D., Sumpton, W., Reid, D., and Mayer, D. G. (2012) Beach areas used by juvenile white sharks, *Carcharodon carcharias*, in eastern Australia. In: Global Perspectives on the Biology and Life History of the White Shark, Domeier, M.L. (ed.). CRC Press, Boca Raton, FL, pp.271–286

West, J. (1996) White shark attacks in Australian waters. In: Great White Sharks: the Biology of *Carcharodon carcharias*, Klimley, A.P., and Ainley, D.G. (eds.), Academic Press, San Diego, CA, pp.449–455

Wetherbee, B.M., Lowe, C.G., and Crow, G.L. (1994) A review of shark control in Hawaii with recommendations for future research. Pacific Science, 48:95–115

White, C.F., Lyons, K., Jorgensen, S.J., O'Sullivan, J., Winkler, C., Weng, K.C., and Lowe, C. G. (2019) Quantifying habitat selection and variability in habitat suitability for juvenile white sharks. PLoS One, 14(5), e0214642

Williams, L.H., Anstett, A., Bach Muñoz, V., Chisholm, J., Fallows, C., Green, J.R., Higuera Rivas, J.E., Skomal, G., Winton, M., and Hammerschlag, N. (2022) Sharks as exfoliators: widespread chafing between marine organisms suggests an unexplored ecological role. Ecology, 103(1), p.e03570

Wintner, S.P., Cliff, G. (1999) Age and growth determination of the white shark, *Carcharodon carcharias*, from the east coast of South Africa. Fishery Bulletin, 97:153–169

Winton, M.V., Sulikowski, J., and Skomal, G.B. (2021) Fine-scale vertical habitat use of white sharks at an emerging aggregation site and implications for public safety. Wildlife Research, 48:345–260

Winton, M.V., Fay, G., and Skomal, G.B. (2023) An open spatial capture-recapture framework for estimating the abundance and seasonal dynamics of white sharks at aggregation sites. Marine Ecology Progress Series, 715:1-25.

ABOUT THE AUTHOR

DR. GREG SKOMAL is an accomplished marine biologist, underwater explorer, photographer, and author. He has been a fisheries scientist with the Massachusetts Division of Marine Fisheries since 1987 and currently heads up the Massachusetts Shark Research Program. He is also adjunct faculty at the University of Massachusetts School for Marine Science and Technology and an adjunct scientist at the Woods Hole Oceanographic Institution. He holds a master's degree from the University of Rhode Island and a Ph.D. from Boston University. For more than 40 years, Greg has been actively involved in the study of life history, ecology, and physiology of sharks.

His shark research has spanned the globe from the frigid waters of the Arctic Circle to coral reefs in the tropical Central Pacific. Much of his current research centers on the use of acoustic telemetry and satellite-based tagging technology to study the ecology and behavior of sharks. Greg has been an avid SCUBA diver and underwater photographer since 1978. He has written dozens of scientific research papers and has appeared in a number of film and television documentaries, including programs for National Geographic, Discovery Channel, BBC, and numerous television networks. His has authored several books, including *The Shark Handbook*. His most recent book, *Chasing Shadows*, chronicles his personal history with white sharks, and the re-emergence of this predator on Cape Cod, Massachusetts. He is a Boston Sea Rover and a member of The Explorers Club; his home and laboratory are on the south coast of Massachusetts.

ABOUT CIDER MILL PRESS BOOK PUBLISHERS

Good ideas ripen with time. From seed to harvest,
Cider Mill Press brings fine reading, information, and
entertainment together between the covers of its creatively
crafted books. Our Cider Mill bears fruit twice a year,
publishing a new crop of titles each spring and fall.

"WHERE GOOD BOOKS ARE READY FOR PRESS"
501 NELSON PLACE
NASHVILLE, TENNESSEE 37214

CIDERMILLPRESS.COM